Fattening Poultry for Meat Purposes

by US Dept. of Agriculture

with an introduction by Jackson Chambers

This work contains material that was originally published in 1911.

This publication is within the Public Domain.

This edition is reprinted for educational purposes
and in accordance with all applicable Federal Laws.

Introduction Copyright 2018 by Jackson Chambers

The World's Largest Selection of Vintage Poultry Books

www.VintagePoultry.com

Self Reliance Books

Get more historic titles on animal and stock breeding, gardening and old fashioned skills by visiting us at:

http://selfreliancebooks.blogspot.com/

Introduction

I am pleased to present yet another title on Poultry.

The work is in the Public Domain and is re-printed here in accordance with Federal Laws.

As with all reprinted books of this age that are intended to perfectly reproduce the original edition, considerable pains and effort had to be undertaken to correct fading and sometimes outright damage to existing proofs of this title. At times, this task is quite monumental, requiring an almost total "rebuilding" of some pages from digital proofs of multiple copies. Despite this, imperfections still sometimes exist in the final proof and may detract from the visual appearance of the text.

I hope you enjoy reading this book as much as I enjoyed making it available to readers again.

Jackson Chambers

LETTER OF TRANSMITTAL.

UNITED STATES DEPARTMENT OF AGRICULTURE,
BUREAU OF ANIMAL INDUSTRY,
Washington, D. C., May 29, 1911.

SIR: I have the honor to transmit the accompanying manuscript entitled "Fattening Poultry," by Alfred R. Lee, of the Animal Husbandry Division of this bureau, and to recommend its publication in the bulletin series of the bureau. The work describes methods of feeding poultry on a large commercial scale, and presents figures on the cost of such feeding. Mr. Lee devoted the greater part of the past summer and fall to this work and collected complete data on the feeding of over 100,000 chickens. He also secured partial data on the feeding of upward of 200,000 others. The results obtained in dealing with these large numbers are unusually important because of the elimination of the errors which are peculiarly liable to occur in drawing conclusions from the feeding of a small or comparatively small number of fowls. The present investigation is, in fact, believed to be the first attempt to acquire comprehensive and reliable figures on the cost of producing a pound of gain in poultry.

The author desires to acknowledge the assistance rendered in the prosecution of the work by Messrs. Harry M. Lamon and C. L. Opperman, of the Animal Husbandry Division.

Respectfully,

A. D. MELVIN,
Chief of Bureau.

Hon. JAMES WILSON,
Secretary of Agriculture.

CONTENTS.

	Page.
Introduction	7
Educating the public taste	8
Conditions in France and England	8
Methods of handling live poultry	9
Best breeds for fattening	11
Comparison of breeds	11
Individuality in chickens	12
The feeding season	14
Length of feeding period	14
Methods of fattening	14
Milk fattening	14
Various methods in vogue	15
Crate or trough fattening	16
Rations	17
Grain mixtures	17
Feather picking resulting from excessive grain feeding	18
Milk or buttermilk essential in all rations	19
The use of tallow	20
Mixing the feed	20
Mixing machines	20
Consistency of the feed	21
Number of times to feed daily	22
Color of milk-fed poultry	23
The feeding stations and their equipment	23
Details of the feeding experiments	31
Experiment A	32
Experiment B	37
Comparison of experiments A and B	42
Experiment C	42
Experiment D	44
Comparison of experiments C and D	46
Average daily consumption of grain per head	46
Daily death records	47
Fattening hens	48
Shrinkage in dressing	50
Cleaning and spraying the batteries	50
Poultry manure	51
Keeping records	52
Conclusions	53
Appendix	55
Table I.—Details of feeding experiment A	55
Table II.—Details of feeding experiment B	58

ILLUSTRATIONS.

PLATES.

	Page.
PLATE I. Fig. 1.—Shipping crate for live poultry. Fig. 2.—Ordinary livestock car, often used for shipping poultry. Fig. 3.—Live-poultry cars, generally used for long-distance hauls.	8
II. Fig. 1.—Portable feeding and mixing tank. Fig. 2.—Portable truck for moving birds. Fig. 3.—Manure truck.	28
III. Fig. 1.—Portable feeding battery—side view. Fig. 2.—Portable feeding battery—end view. Fig. 3.—Turkey-feeding battery. Fig. 4.—Two types of feed pails.	28
IV. Fig. 1.—Feeding station No. 2. Fig. 2.—Feeding station No. 3. Fig. 3.—Feeding station No. 4. Fig. 4.—Combination creamery and poultry-feeding station (station No. 5).	28

TEXT FIGURES.

FIGURE 1.—Stationary feeding battery—end view	24
2.—Stationary feeding battery—front view	25

FATTENING POULTRY.

INTRODUCTION.

An opportunity for obtaining extensive and accurate data of the results secured in fattening poultry under commercial conditions in poultry-packing houses, which came up during the feeding season of 1910, was gladly made use of by the author, with the hope of throwing some light upon the cost of fattening poultry commercially, and also to show the relative gains which one can expect to secure with the different grades of poultry. Through the courtesy of two companies operating large poultry-packing houses in the Middle West it was possible to compare the results secured at different feeding stations and to study the efficiency of various methods, as well as the feeding value of different rations. The author is especially indebted to these poultry packers, who extended every possible courtesy to him and gave him free access to all the details of that part of their business in which he was interested.

The methods described are the result of years of experience and extensive, practical experimentation by these packers. As most of the experimental work hitherto published on poultry problems deals with small numbers, this careful study of a large number of birds suggested itself as a good opportunity to show what influence different factors may have in causing variation in results. The results of the season's work fulfilled the author's hopes to a considerable degree. The study of poultry problems under successful commercial conditions permits an investigation of the vital problems of the work without encountering the many difficulties which are liable to be met in starting a poultry plant for experimental purposes. This method also involves the use of large numbers of birds, thus eliminating many errors which may occur in drawing conclusions from small numbers.

Whatever value this work may have, aside from the comparative experimental data, lies in the fact that it describes in detail successful methods of fattening chickens for market and shows what was the cost of producing the gains. This work has been carried out successfully on a commercial scale in the Middle West, and as there is a tendency for the extension of poultry-packing houses southward, as well as a decided growth of the industry generally in the South, material on this subject should be of special value in that section. The normal growth of poultry interests in the Middle West develops a corresponding growth in the poultry packing and feeding business. Numerous experiments have been conducted showing the cost of producing a pound of gain in fattening steers, hogs, and sheep, but very little work has been published showing the cost of producing gain in poultry. The experimental work which has been published on fattening poultry has been of value rather as showing the comparative value of rations than the average cost of producing poultry flesh, as in many experiments the actual cost of producing the gains has been so high that it would not be commercially profitable.

In order to produce a superior quality of chicken flesh for high-class eastern and foreign buyers, managers of the poultry and egg packing houses of the Middle West for a number of years have fed "spring" chickens, producing the so-called "milk-fed" chickens. If these chickens were fed and fattened on the farm, it would not be necessary for the packer to put them into a feeding station, but the bulk of the chickens produced on the farm are too thin to make first-class dressed poultry without special fattening.

If the farmer gave his chickens a daily supply of grain, he could, in many cases, fatten his chickens at a profit before sending them to market, even when he sells to the large poultry buyer. He, of course, could not afford to put in an expensive equipment for fattening, but he could get his chickens in fair condition by supplying grain daily, or by confining those to be marketed for two or three weeks, and feeding freely either on corn, or on corn meal, wheat flour, or oat flour, mixed with skim milk or buttermilk.

EDUCATING THE PUBLIC TASTE.

When a bird has been properly fattened oil replaces much of the water in the flesh, so that when it is cooked the flesh becomes tender and juicy. Many consumers of poultry do not know how delicious a well-fattened spring chicken is, but after once securing a bird thus fattened they will most likely ask for the same quality in the future. As the people of this country become acquainted with the taste of chicken of good quality the demand will grow and they will be only too willing to pay for the extra cost of well-fattened birds. Most of the dressed spring chickens found in the average market to-day have been insufficiently fed; they can not be classed as fat.

CONDITIONS IN FRANCE AND ENGLAND.

The people of France and England have appreciated for some time the value of properly fattened poultry. Various methods of fattening are in use in both of these countries. In France most of the poultry is fattened on the farms where it is raised by the farmer, who is usually very skillful in this art. Much of this fattening is done by hand, involving more labor than the American farmer can afford to give for this purpose, both on account of the higher cost of labor and of the lower price paid for the finished product in this country. Most of the special fattening in England is done at large establishments where the birds are confined in crates and fed from troughs for 7 to 10 days and then finished with cramming machines, making the total fattening period about 3 weeks. The coops or "batteries" in which the birds are fed are often placed out of doors, with some protection from the wind and rain. During cold weather the batteries are put into a building in order to conserve some of the body heat of the chickens; otherwise much of the value of the feed would be consumed in keeping the body warm rather than in producing flesh. Besides the extra care in feeding, special attention is given to selection and breeding, so as to build up strains which fatten readily. Much care is also taken in dressing the poultry, so that it is offered to the public in a very attractive condition. The farmers and special poultry keepers in this country could with good advantage adapt some of these methods to their own conditions.

METHODS OF HANDLING LIVE POULTRY.

Poultry buyers located in the towns and villages of the Middle West receive chickens from the farmers throughout the greater part of each year, but the stock used in commercial fattening is shipped in from June until the following December or January, the feeding season beginning earliest in the southern part of this section. The chickens are handled in many ways, often coming from the farm tied in lots of 4 to 6, with a tight cord around their legs or piled in burlap sacks, so that either their legs are scraped raw or become numb or they are half suffocated when they reach the local buyer or country merchant, who sells to the poultry buyer. It is difficult to understand the cruel and careless methods of some persons in handling their poultry, when by using a little care they could send their poultry to the buyer in good condition and save themselves money. The following styles of crates are used extensively by the packer and poultry buyer and could easily be adapted for use on the farm at a small cost and result in a great saving in labor as well as improving the condition of the poultry.

The coop shown in Plate I (fig. 1) is a very good type of wire crate, simple in construction and cheap. This crate is 2 feet 4 inches wide, 3 feet 10 inches long, and 13 inches deep, inside measurements, and weighs 28 pounds. The uprights are of furring $\frac{7}{8}$ by $1\frac{3}{4}$ inches; the floor is solid and made of $\frac{1}{2}$-inch boards and the sides and top covered with strips of 2-inch mesh wire, 1 foot wide, with a partition of similar material dividing the crate into two equal parts to prevent the birds from all bunching together. Shipping crates of about this size made entirely of wood weigh from 32 to 36 pounds, depending on the width of the wooden slats and the size of the corner posts, but they are no better than the wire crate and materially increase the weight, and consequently the cost of shipments. Wire crates of the type illustrated are also used for shipping turkeys, but are made 19 inches instead of 13 inches deep. Coops are also made of galvanized-iron strips and wire and of all kinds of modifications between those made entirely of wire and those made wholly of wood.

Some styles of crates have square wire doors on the top, while others have a thin slat which slides under three narrow strips of tin or iron and is fastened in the center with one nail. The coop illustrated is opened by moving a slat, fastened by a spring, which forces a small cut in the slat against a wire in the center. A coop thus shut can be opened easily and quickly, but occasionally gets out of order. There is a tendency to break the slats of coops which are nailed when opening them at the packing houses, especially if the work is done in a hurry.

The styles of coops above described are used in shipping either by express or by freight. The live poultry is often held for a day by the local poultry buyer and then shipped, either by express or freight, to the poultry and egg packer. If shipped by freight, an open car ordinarily used for shipping live stock locally is devoted entirely to eggs and poultry, which are picked up at each station and piled into the car, poultry generally at one end and eggs at the other. (See Plate I, fig. 2.) This car on arrival at the packing house is unloaded immediately if it comes in during regular working hours; if not, it is left till morning, although the eggs are often unloaded by the night force.

The birds are fed by the small poultry buyers if held for any length of time, and grain is scattered in the coops before they are shipped to prevent a heavy shrinkage in weight. The chickens are usually hungry by the time they are distributed in the feeding station and are not held long without feed. Chickens shipped in by express generally have less feed in their crops than those shipped by freight.

Poultry is also shipped extensively in cars built for live poultry (Plate I, fig. 3), especially when their destination is far enough away so that they will be over a day on the road. These cars, which are used extensively in interstate shipments, are of the following dimensions: 36 feet long, 9 feet 5 inches wide, inside measurements, with a "stateroom" 8 feet by 9 feet 6 inches in the center of each car and an aisle 2 feet 3 inches wide extending the entire length of the car. Each car has 8 tiers, which gives about 1,600 square feet of coop floor capacity, and each tier accommodates 16 coops. Allowing 36 fowls to each coop, cars of these dimensions will accommodate about 4,600 head, or about 18,000 pounds, of poultry, depending on the average size of the birds. About 2,000 to 2,400 geese or 1,200 to 1,500 turkeys make a carload. A water tank of 327 gallons capacity and a grain crib 8 feet square and 20 inches deep are attached to each car. All compartments have feed and water troughs accessible from the aisle, in which rations are fed consisting of corn meal, corn chop, and a small per cent of shorts in different proportions mixed with buttermilk. Dead birds can be easily seen and readily removed from all coops. These cars are well ventilated and carry the birds to their destination in good condition, the shrinkage rarely exceeding 5 per cent.

BEST BREEDS FOR FATTENING.

All varieties and types of chickens are fattened in this country, no special attention being devoted to developing strains or special types peculiarly adapted to produce a high quality of flesh or to give especially good results in fattening. Several breeds give good results in fattening, and these are preferred by men who make a specialty of feeding poultry in the following order: Plymouth Rocks, Wyandottes, Rhode Island Reds, or taken as a whole, birds of the general-purpose breeds. The feeders discourage the use of birds of the Mediterranean class, such as the Brown and White Leghorns and the Minorcas, because these birds average poorer results throughout the season in the feeding tests and they mature light, while the trade demands a heavy fowl. In order to make the farmer raise chickens of the first rather than of the second class mentioned, the packer, and consequently the smaller poultry buyer, often pays from 1 to 3 cents a pound less for light-weight hens.

More attention should be paid to buying poultry on a quality basis, thus showing the producer the gain which he may realize by keeping good poultry of the general-purpose breeds and giving the birds proper attention and feed before shipping to market. In some sections the packers have exchanged purebred cockerels of the general-purpose breeds for the mongrel and light-weight cocks kept by the farmers, thus rapidly improving the quality of stock in the localities where they obtain their supplies. The Orpingtons, various game crosses, and the Dorking make good poultry for fattening, but are not found in any appreciable numbers in the Middle West, although the Orpingtons have increased considerably in the last few years.

COMPARISON OF BREEDS.

Table 1 gives the gains secured in selected "batteries" in which the birds were sorted in feeding experiment A (see p. 32). It may here be stated that the coops in which the birds are kept during the feeding period are called batteries. There are two styles of batteries—stationary and portable—most feeding houses containing the stationary batteries. The per cent of gains shows that in general Leghorns make much poorer gains than Plymouth Rocks, but the results are not entirely consistent. The Leghorns make poor gains in the batteries after they are 2½ or 3 months old, as they are very restless, but they make good gains up to this age. Leghorns mixed with other birds in the compartments of the batteries tend to keep all of the birds excited. The chickens in Table 1 were all fed between September 22 and November 14.

TABLE 1.—*Results of fattening various breeds of chickens.*

Number of birds.	Breeding.			Days fed.	Average weight in.	Average weight out.	Died.	Average gain.	Gain.
	Rocks.	Leghorns.	Mixed.						
	Per cent.	*Per cent.*	*Per cent.*		*Pounds.*	*Pounds.*		*Pounds.*	*Per cent.*
68	100			14	3.68	4.25		0.57	15.5
192	100			14	4.03	4.82		.79	19.6
80	100			14	2.85	3.70		.85	29.8
80	100			9	4.01	4.61		.60	15.6
80	100			9	2.38	3.00		.62	26.1
80	100			7	3.26	3.78		.52	15.9
80	92	8		10	2.83	3.56		.73	25.8
80	87		13	8	3.18	3.85		.67	21.1
80	33		67	9	2.98	3.36		.38	12.8
80		100		9	2.88	3.19	1	.31	14.8
80		80	20	9	2.29	2.87	1	.58	25.3
80		65	35	6	2.16	2.55		.39	18.1
80	24	60	16	8	3.60	3.93		.33	9.2

INDIVIDUALITY IN CHICKENS.

Table 2 gives in detail the gains of the batteries making up two of the lots in feeding experiment A (p. 32). The first 11 batteries went on feed September 18 and the remaining 8 September 28. Records were kept of the gains of each battery. A lot includes all batteries put on feed on the same day and fed the same length of time. The batteries in the table are arranged in the relative order of the increasing average weights of the birds, but the percentage of gain does not vary directly with the average weight. The "Rocks" in this table were birds of the general-purpose classes, Barred Plymouth Rocks predominating. If the batteries were arranged in order according to the proportion of Leghorns the per cent of gains would still show no consistent relative order. These gains show a more consistent ratio between the average weight of the birds and the per cent of gain, which in general varies inversely. If the average weight is lowered by having a large proportion of Leghorns in the battery, this inverse ratio is not so apparent. This table shows the great variation in lots housed and fed alike, and emphasizes the great difference in the ability of the individual bird to put on flesh. This difference is greater because of the mixed stock in each battery. All the other lots in feeding table experiment A could be subdivided, and would show a similar variation in gains, but these two lots were selected as showing about the average variation within a lot. This emphasizes the error which is likely to occur in experimental work dealing with small lots, unless the chickens are of the same strain and have been handled alike from birth.

TABLE 2.—*Individuality in chickens.*

Number of birds.	Breeding.			Days fed.	Average weight in.	Average weight out.	Died.	Average gain.	Average gain.
	Leghorns.	Rocks.	Mixed.						
	Per cent.	Per cent.	Per cent.		Pounds.	Pounds.		Pounds.	Per cent.
80	10	90		8	1.89	2.24		0.35	18.5
80	55	45		8	1.98	2.41	1	.43	21.7
80	21	79		8	2.10	2.42	1	.32	15.2
80	35	65		8	2.13	2.33		.20	9.4
80	34	66		8	2.25	2.69		.44	19.6
80	16	84		8	2.31	2.83		.52	22.5
80	15	85		8	2.44	2.98		.54	22.1
80	9	91		8	2.46	3.05	1	.59	24.0
80	12	88		8	2.53	2.95		.42	16.6
80	12	88		8	2.70	3.09		.39	14.4
80	5	95		8	2.84	3.03	2	.19	6.7
80		65	35	6	1.60	2.20		.60	37.5
80	8	75	17	6	2.28	2.85		.57	25.0
80	29	13	58	6	2.35	2.83		.48	20.4
80	5	70	25	6	2.49	2.90		.41	16.5
80	10	58	32	6	2.50	2.81	1	.31	12.4
80	16	41	43	6	2.59	2.98	1	.39	15.1
80		58	42	6	2.59	3.00	1	.41	15.8
80	5	25	70	6	2.78	3.32	1	.54	19.4

The "feeder" who has charge of the fattening station often finds some in each lot of chickens that it would not pay to fatten. In such cases the Leghorns (especially the Single Comb Brown), Buff Cochins, Langshans, and all chickens with black legs are discarded. Varieties containing much Buff Cochin blood are claimed by some feeders to be very poor varieties to fatten. In sorting the birds the feeder sometimes throws out many birds in good market condition which it does not pay to feed under his conditions, as in many instances they would lose rather than gain in weight during the feeding period. If one can select the birds still more carefully for fattening he should pick out birds with short, stout, well-curved beaks, broad heads, bright, clear eyes, deep, broad breasts, and well-spread legs. Individual birds of the same breed vary greatly in their ability to put on flesh, but a strain can be selected and bred for this purpose. The superiority of certain strains of birds in England and France, in their ability to fatten readily, is quite marked. Crosses are frequently used in England for producing good fattening stock, but the majority of farmers and poultrymen in this country do not breed carefully and systematically enough to get good results from crossing, so that the offspring show lack of uniformity in type and size, which tends to lower the price paid for the birds. It is probable that feeders could with profit select birds along some of the lines mentioned in this paragraph, but this matter depends largely on each man's condition.

THE FEEDING SEASON.

The feeding season varies in different sections, depending on the climatic and seasonal conditions, and on the trade supplied with the dressed poultry. The season begins earlier in the South, generally in June, and lasts longer in the North, up to January or February, depending on the supply of live chickens. Many find it quite profitable to feed chickens as broilers during the early summer months, while other feeders prefer to feed the larger sizes, called "springs" and "roasters."

LENGTH OF FEEDING PERIOD.

The common practice in poultry packing houses is to feed each lot 17 days or less. The market or trade supplied and the results secured by the feeder determine the length of the feeding period. Most milk-fed chickens are fed for 14 days, but results secured in feeding indicate that a more profitable gain can be secured in a shorter feeding period, provided the same price per pound can be secured for the finished product. In England and Canada birds are fattened for at least 3 weeks, and if one uses a cramming machine it probably pays to feed for that length of time. If the birds are small or thin, they may be fed longer than heavier birds or those which are fairly well fleshed when they reach the feeding station. As the feeding season advances the tendency among feeders is to shorten the length of the feeding period, reducing it as low as 7 days in many cases. Many birds are merely "finished" by feeding for 5 to 6 days, and these are not generally classed as milk-fed poultry.

METHODS OF FATTENING.

MILK FATTENING.

Practically all of the special feeding in this country involves the use of milk, thus producing "milk-fed" chickens. These are also exported to some extent. Milk, while the least expensive, seems to be the most essential constituent of the ration, and when a feeder can not get milk in some form he generally does not attempt to fatten poultry commercially. The profit depends on various factors, many of which are local, and must be worked out by each individual. Among these factors are the supply and cost of the chickens, which depends largely on the competition of other buyers; the shipping facilities; the cost of the essential feeds; the availability and cost of efficient labor; the market, and the price which the packer can secure for his finished product. Often the packer has to feed his poultry to suit the demands of his market, but generally if a man has a high-class product he can make his own market, catering somewhat to popular fancies.

METHODS OF FATTENING.

Besides these local factors there are certain essentials to success in a feeding station where poultry are fattened. First in importance is the manager of the station, or feeder, who must thoroughly understand all the details of the work and have a well-trained, observant eye, quick to note the condition and appetite of the stock. Success or failure depends primarily on this man, who must have the knack of caring for birds. The feeding station must be arranged to economize labor and to provide the best possible ventilation. Conditions must be of such a nature as to keep the birds quiet and contented, and at the same time cause them to consume a large amount of feed, in order to make profitable gains.

VARIOUS METHODS IN VOGUE.

Besides crate fattening from troughs there are several other methods in vogue, particularly in Europe. Among these are fattening by funnel, by machine, and by hand. The last is common in France, but can only be done economically where labor is cheap. The funnel method is used to some extent in England and France, with the funnel tube running directly to the crop, which is filled by pouring the mixture into the funnel. The other method, cramming by machine, is used extensively in England, generally to supplement trough feeding. The English feeder does not consider that the bird has been properly fattened until it has been finished with a cramming machine. Most of the large feeders have used cramming machines in the United States, but have not found them adapted to their conditions. There are two factors which may help to account for this attitude: First, very few feeders in this country have been able to use a cramming machine successfully and keep the birds contented; and, second, the trade has not been educated to the increased value of a machine-fed bird. However, the method is occasionally found in use where there is a special market for birds which have been crammed.

Some feeders in this country have obtained good results with the machine in one section, and made an absolute failure of the same method under different conditions. In England the art of fattening by machine is often handed down from father to son, thus producing first-class feeders. The cramming machine is used to some extent in this country for fattening hens which do not give good results on trough feeding.

In cramming, the birds are fed from 7 to 14 days from the troughs, and are then crammed twice daily for from 7 to 10 days, until they begin to go off feed, when they are marketed. The operator gauges the proper amount of feed to force into the birds by holding his hand on the bird's crop. If the crop is not almost or entirely empty at the next feeding time the bird is not given any additional feed.

Another method which is used to a considerable extent on a small scale in this country is pen fattening. This method is adapted for use on the farm, where the farmer does not care to go to the trouble of crate fattening, or where the price received for well-fed birds does not warrant the extra labor and feed cost of the latter method. Pen fattening has in some cases given very good results, but it is not as reliable as crate fattening, although the labor cost is less. It is used generally in fattening ducks. The quality of flesh secured by crate fattening is better than that obtained by pen fattening.

CRATE OR TROUGH FEEDING.

Crate fattening from troughs is the method of feeding employed in this country by most of the large fattening establishments. From 6 to 10 chickens are placed in the crate or battery, generally with a small amount of feed in their crops, and given a light feed at the next regular feeding period. Two methods are used in transferring the chickens from the coops to the feeding battery. The coops of chickens are weighed on scales located at some convenient place on the dock or in the packing-house building, and then put either into a portable transfer crate or directly into the portable feeding battery. If the birds are put directly into the portable battery, it saves the labor of rehandling and they go on feed in better condition than if rehandled. Many birds are graded into a lower class on account of broken wings, sometimes caused by handling after the birds have been on feed for some time; thus the use of the portable feeding battery tends to lessen the loss caused by rehandling. To facilitate the weighing of coops two strips of wood are nailed onto the scale platform, thus elevating the coops so that they do not touch the floor on any side, and making it easier to handle them. Some people advise dusting the birds before putting them into the batteries, but by keeping the batteries and coops clean and whitewashing frequently, the large feeders find that it is not necessary in the case of short feeding periods to dust the birds with lice powder.

From 6 to 10 birds are placed in each division of the battery, depending on the size of the birds and the ideas of the feeder. Ten birds seem rather a large number to place together, but very good results have been secured with this number in the portable feeding battery hereinafter described, although it would seem advisable to reduce the number to 8 when the birds weigh from $3\frac{1}{2}$ to 4 pounds. Two or three chickens do better in a division together than when only 1 bird is placed in each compartment, and the cost of equipment and labor per bird varies inversely with the number of birds in each division.

RATIONS.

A perusal of the literature on the subject of chicken feeding indicates that there is a large variety of grain feeds, mixed in varying proportions, which are successfully used in fattening. But many of the large poultry feeders after trying various feeds and rations have found that a very simple ration, made up of only two or three grains, is best suited to economical gains under their conditions. Considering the large number of birds which they feed each year and the extent of their experimentation in feeding, it would appear that these simple rations must be of special merit for their conditions. Most rations are recommended for a feeding period of three weeks, although the length of the feeding period may influence the selection of the best ration. Birds fed only for a short time may be forced on highly concentrated feeds, whereas birds fed longer may need a ration containing a greater variety and less concentrated. While this may be true, many of the poultry packers feed the same ration to their chickens regardless of the length of the feeding period. The fact that it is easier to feed only one mixture may help to explain this condition, or it may be possible that the ration is not too heavy or concentrated even for the longer feeding periods.

GRAIN MIXTURES.

In selecting a ration the feeder must be influenced to some extent by the market price and supply of grains. Certain grains which are used for fattening are peculiarly adapted to local sections, and are not widely distributed at reasonable prices on the general markets. Among such grains are buckwheat, pea meal, graham flour, shredded-wheat waste, small potatoes, and in some places barley meal.

In the feeding records given in this bulletin the following rations were used: No. 1, 60 per cent corn meal and 40 per cent low-grade wheat flour; No. 2, 58 per cent corn meal, 36 per cent oat flour, and 6 per cent tallow, by weight. Ration No. 2 was varied during the season to suit the fancy of the feeder or the changes in the weather. This variation of the ration was not regular, but, generally speaking, as the season advanced and the weather became cooler a larger proportion of corn meal was fed, although the increase was not large if figured on the average per cent of corn meal in the ration each month. These two rations were selected by different individuals working under slightly different conditions. In general they are quite similar, except that tallow is present in one ration and not in the other. Many feeders after experimenting with a large number of different feeds have returned to these simple rations.

There are many other rations which have been used with good results, and perhaps are specially suited to certain localities on account of the relative price of grains. The following rations are mentioned as belonging to this class: No. 3, 2 parts of oat flour, 1 part of barley meal, and 1 part of corn meal; No. 4, 2 parts of oat flour, 1 part of barley meal, and 1 part of boiled potatoes; No. 5, 1 part of corn meal, 1 part of oat flour, and 1 part of wheat flour; No. 6, 2 parts of corn meal, 2 parts of buckwheat flour, and 1 part of ground oats; and No. 7, 1 part of oat meal, 1 part of graham flour, and 1 part of corn meal, by weight.

In France and England buckwheat flour, oat flour, and barley meal are used extensively in fattening. Shredded wheat waste has been used to replace oat or wheat flour, with good results, in places where it could be bought at a low price.

These rations include most, if not all, of the grains which are used extensively in this country, but there are many other combinations which have been, and are still, used for fattening.

All feeders are very particular that the grains used are of the best quality, and they find it especially necessary to watch the oat flour, often returning a shipment as unfit for their use. The oat flour, with the hulls removed, must be finely ground and should give a sweet taste when chewed. In some cases feeders have been forced to substitute other kinds of flour for the oat flour, as they could not always secure good quality oat flour, which is apt to contain other grains.

Grit is generally provided if the birds are kept on feed for two weeks or longer, giving 4 pounds of grit to 100 birds if fed twice a week. Where the birds are only fed a short time, 7 to 10 days, they do not need grit if they were raised in a section sufficiently supplied.

Clover or alfalfa meal, meat meal, blood meal, charcoal, and salt are often added to the rations, according to the fancy of the feeder. These constituents do not appear to be absolutely essential, but may be worth while for special conditions.

FEATHER PICKING RESULTING FROM EXCESSIVE GRAIN FEEDING.

Birds often become very restless on forced feeding of a highly concentrated ration, and commence feather eating and picking each other, often continuing until they have eaten a considerable part of the flesh of a live chicken. Probably an overheated condition of the blood, caused by consuming a large proportion of highly heating feed, such as corn meal, during hot weather will lead to habits of this kind. In such cases it may be advisable to reduce the proportion of corn meal and lighten the ration by adding some green feed, such as clover or alfalfa meal, and possibly a small amount of meat or blood meal. Salt, sulphur, or powdered borax, lightly sprinkled into the mash, have been suggested as remedies for this overheated condition of the blood. Salt is quite frequently used, but sulphur and borax, if fed in any appreciable quantity, appear to lessen slightly the appetite of the birds, although the difference is not marked. It is rather difficult to prove what effect these substances have in lessening the chance of the birds developing these bad habits; in any event these remedies are probably not used extensively.

MILK OR BUTTERMILK ESSENTIAL IN ALL RATIONS.

Milk is used entirely in mixing the various rations used in fattening, and is considered an essential ingredient, both in this country and in Europe. While good results may be secured without it, milk has such a beneficial effect on the birds that it is hard to get good results without using it. In some instances poultry shippers stopped feeding chickens when their supply of milk gave out. Buttermilk and skim milk are generally used, no particular notice being taken as to whether the milk is sweet or sour, but in almost all cases it was sour before it was fed to the birds. One large creamery in Kansas has experimented in condensing buttermilk, and they now manufacture a product which is put up in barrels, stored, and sold for feeding chickens. In one case they shipped carload lots to a feeding station located several hundred miles away. As many creameries have a large surplus of buttermilk during the spring months, this appears to be quite a profitable way of disposing of it, except that it involves expensive machinery, and that the condensed milk, being bulky, requires a large amount of storage space. This creamery was unable to satisfy half of the demand for condensed milk during the fall and early winter months. The milk is reduced to one-fourth of its original volume largely by evaporation, but part of the whey is drained off during this process. When condensed the milk is run into barrels without adding any preservative, and will keep indefinitely. Sample barrels have been kept for two or three years, and when opened the contents have been in good condition. Under ordinary trade conditions the milk would never be kept longer than one year. Fresh buttermilk, condensed buttermilk, and skim milk are preferred in this relative order. Whey is sometimes fed in addition to the condensed buttermilk, but it is too bulky and of too small feeding value to pay to move any considerable distance.

Various feeders have endeavored to find a substitute for milk, with little apparent success. Milk seems to have a very important influence on the digestive processes, keeping the bird in good condition under forced feeding. Beef broth has been used to some extent, with fair results, but it is not as good as milk. If the feed is mixed with water, from 5 to 15 per cent of the ration should be meat in some form, and vegetables or green feed should be added. Green feed is fed to some extent in very hot weather, but most feeders do not think that the results warrant its use. Finely ground beef scrap and meat meal are good forms of meat feed.

THE USE OF TALLOW.

Beef tallow is used by many feeders, but has been discarded by others, who claim that it produces a poorer quality of flesh. When only a very small amount is fed the difference in the flesh is not noticeable; but, considering the cost of the tallow and the possible poorer quality of flesh produced, it hardly seems to be an economical feed, although this depends largely on individual conditions, especially on the market to which the packer sells. In part of the feeding experiments in this bulletin about 6 per cent of the ration, excluding milk, consisted of tallow, and this had no apparent effect on the flesh. Tallow is often recommended to be fed during the last few days of the feeding period, but under ordinary commercial conditions it is hardly practicable to mix the feed separately and use it according to the number of days which the birds have been in the feeder. The tallow may be shaved directly into the feeding trough, but this method does not seem as practicable as to mix the melted tallow into the feed.

MIXING THE FEED.

The feed may be mixed with a rake or in a machine; some feeders preferring to mix with the rake regardless of the amount which has to be mixed. The feed can be mixed fairly quickly with a rake by a skillful feeder, but most feeders prefer to let a machine do the mixing where a large number of birds are fed. Some kind of power is necessary to run the mixer. When mixed by an iron rake the milk is run or poured into a large mixing tank and the grain added gradually, constantly stirring the mixture with the rake to prevent the formation of lumps, and to mix the different grains thoroughly. The feeder adds the different grains alternately, generally dumping in 100 pounds at a time, and mixing is continued until the mixture is of an even consistency. It is very necessary to have the feed free from lumps. Tallow may be kept in an open kettle heated by steam pipes, and gradually added to the feed in a melted state, after the milk and grain have been mixed. The pail in which the tallow is handled should be heated before it is used for the melted tallow, to prevent the liquid from congealing on the sides of the pail. The tallow is stirred thoroughly into the mixed feed.

MIXING MACHINES.

There are several styles of machines used for mixing the feed, each manager having his own ideas of the best kind of mixer. A horizontal mixer made of 2 tanks each 6 feet 6 inches long, 2 feet 6 inches wide at the top, and 2 feet 9 inches deep, containing a dasher running lengthwise of each tank, was used at station No. 4 with good results. There were 22 paddles on the horizontal shaft or dasher, set at different angles, each 13½ inches long, 1¼ inches wide, and three-eighths of an inch thick. When the machine was going these paddles barely missed the sides and bottom of the tank, which was concave on the bottom. The narrow side of the paddles cut the feed when in motion. The feed was held and mixed on a platform 15 feet by 6 feet, which was level with the top of the mixing tanks and was built flush against the tanks. The machine was made up of two sections, so that either one could be used separately. A single mixing machine built along similar lines, but upright instead of horizontal, was used at station No. 2, but this did not give very satisfactory results as the consistency of the feed varied when run out at the bottom of the tank. This tank had only one spout or valve, but later a second valve was inserted, which saved considerable time in filling the feeding pails. Each tank at station No. 4 had two valves for letting out the feed. A smaller mixing machine made in the form of a drum was used at two feeding stations not described in this bulletin. These drums opened on the side and could be dumped quickly, but their capacity was small. All of the tanks and mixing machines are fitted with steam pipes so that in cold weather the feed can be heated and fed while warm. If the feed is to be fed warm the milk is heated before adding the meal and flour.

CONSISTENCY OF THE FEED.

The feed is mixed to the consistency of thick cream, or so that it will drip from the tip of a wooden spoon. In very hot weather it is advisable to mix the feed thinner than in cooler weather, and results appear to indicate that one feed daily of a thin mixture with one or two thicker feeds makes the best feeding plan, although opinions differ on this point. The chickens seem to prefer the thicker feed, but it is apt to satisfy their appetites before they have consumed as much feed as they would if the mixture was thinner. This matter has to be left largely to the judgment of the feeder, but it should be observed carefully. As the birds receive no liquid except what they get in their feed, it is necessary to use quite a large proportion of milk in the feed. The percentage of milk used seems to depend on the kind of grains in the mixture, on the weather, and on the feeder. It varies from 55 to 70 per cent, and an average of 60 per cent or a trifle higher seems to give very good results.

The successful feeding of poultry depends largely on the ability of the feeder to notice the condition of the chickens on feed. Birds should be fed lightly for the first two or three feeds, gradually increasing the amount until they receive all they will eat up clean. The feed is poured into the troughs by the feeder, who walks rapidly through the aisles between the batteries, feeding a large number of birds in a short time. The condition of the birds when they go into the battery and the length of the feeding period have considerable influence on how soon to feed the birds the maximum amount. Ordinarily the birds are rather hungry when they go into the batteries, especially if they have been shipped in by express, and they can be fed quite freely from the first feed. If they have feed in their crops when put into the batteries, it is usual to feed very lightly for two or three feeds until they are quite hungry and have become accustomed to their new surroundings. Observations made on a considerable number of birds fed within a short time after they were put into the batteries showed that this practice was a good one under certain conditions, and that in many instances it was not advisable to feed a light ration as long as is ordinarily advised in fattening chickens. The main object in feeding should be to keep the birds' appetites keen and at the same time make them eat as much feed as they can assimilate.

NUMBER OF TIMES TO FEED DAILY.

Birds are fed from two to five times a day, but the more common practice is to feed either two or three times. A skillful feeder can get good results feeding twice daily and many prefer this method; but excellent results are secured by feeding three times a day, even by those who are not experienced feeders. An inexperienced person is apt to get better results by feeding three times a day rather than twice. Regular feeding is necessary, and if the birds are fed twice daily the intervals between the feeding times should be as nearly equal as possible. In this case it is well to feed at 6.30 a. m. and 3 p. m. If the birds are fed three times, feed at 6.30 a. m., 12 noon, and 4 p. m. The feeding hours must be regulated somewhat by the season of the year, by the appetite of the birds, and by the hours which the men are employed. By feeding a small amount often, the birds can be made to eat a larger quantity and their appetites kept keener. Each feeder must decide for himself whether there is enough to be gained by feeding oftener to pay for the extra labor involved.

A good many birds die when on feed, especially during certain seasons. The loss is greatest during hot summer weather, when the birds become prostrated with the heat; and later during October and November, when many of the birds develop some form of sickness. The batteries must be examined closely every day and sometimes twice daily, and all the dead or sick birds removed. While making the rounds for dead and sick birds some feeders find that other birds which are healthy but off feed may be removed and dressed at once, instead of keeping them on feed and having them lose in weight, or possibly become weak and sickly.

COLOR OF MILK-FED POULTRY.

A bleached appearance is very characteristic of milk-fed chickens. Milk is apparently the chief factor in causing this appearance, although the composition of the ration doubtless affects this point to some extent. The birds which were fed 14 days showed the effect of bleaching very plainly and a large proportion were white. Some, however, did not appear to bleach at all. The color in the lots fed from 6 to 9 days was more uneven, the birds showing streaks of yellow and white, although many were fairly white and even in color. Records kept of the comparative number of white and yellow birds at various intervals during the season in experiments A and B showed that in the former 73 per cent were white and 27 per cent yellow, while in the latter only 59 per cent were white and 41 per cent yellow. This would indicate that the use of low-grade wheat flour produced a whiter flesh than the oat flour, but there was considerably more milk in the ration of experiment A, and the larger proportion of milk may have influenced the color of the flesh more than the kind of grain. Allowing for this difference in milk it appears that the wheat flour tends to whiten the skin and flesh as much, if not more, than the oat flour. These records are not strictly comparable, as the dressed birds in experiment A, while uneven in color, were classed as white if they showed the effect of bleaching to any considerable extent. Butter color is sometimes added to the feed to give a rich yellow color to the flesh, but this was only done at one of the feeding stations in this bulletin, and only to a very few lots. Molasses was used for coloring during the season of 1909 at the same station and was said to have given a deep-yellow product. The packers stated that they did not care whether the dressed poultry was white or yellow except in a few lots where the market wanted a certain colored flesh.

THE FEEDING STATIONS AND THEIR EQUIPMENT.

Feeding stations, which are buildings used entirely or principally for fattening chickens, are operated in connection with most large poultry and egg packing houses, and are practicable in all large poultry producing sections where the farmer sells his poultry in relatively poor condition, provided the facilities for shipping or marketing are such that poultry can be held and shipped under cold storage conditions. These stations are generally located at or near a railroad junction or center in order that supplies may be drawn from a large territory. There are many different types of feeding stations, but all should be constructed to economize labor, to provide room for a large number of birds per square foot of floor space, and to keep the birds contented and healthy. The weakest point in most feeding stations is

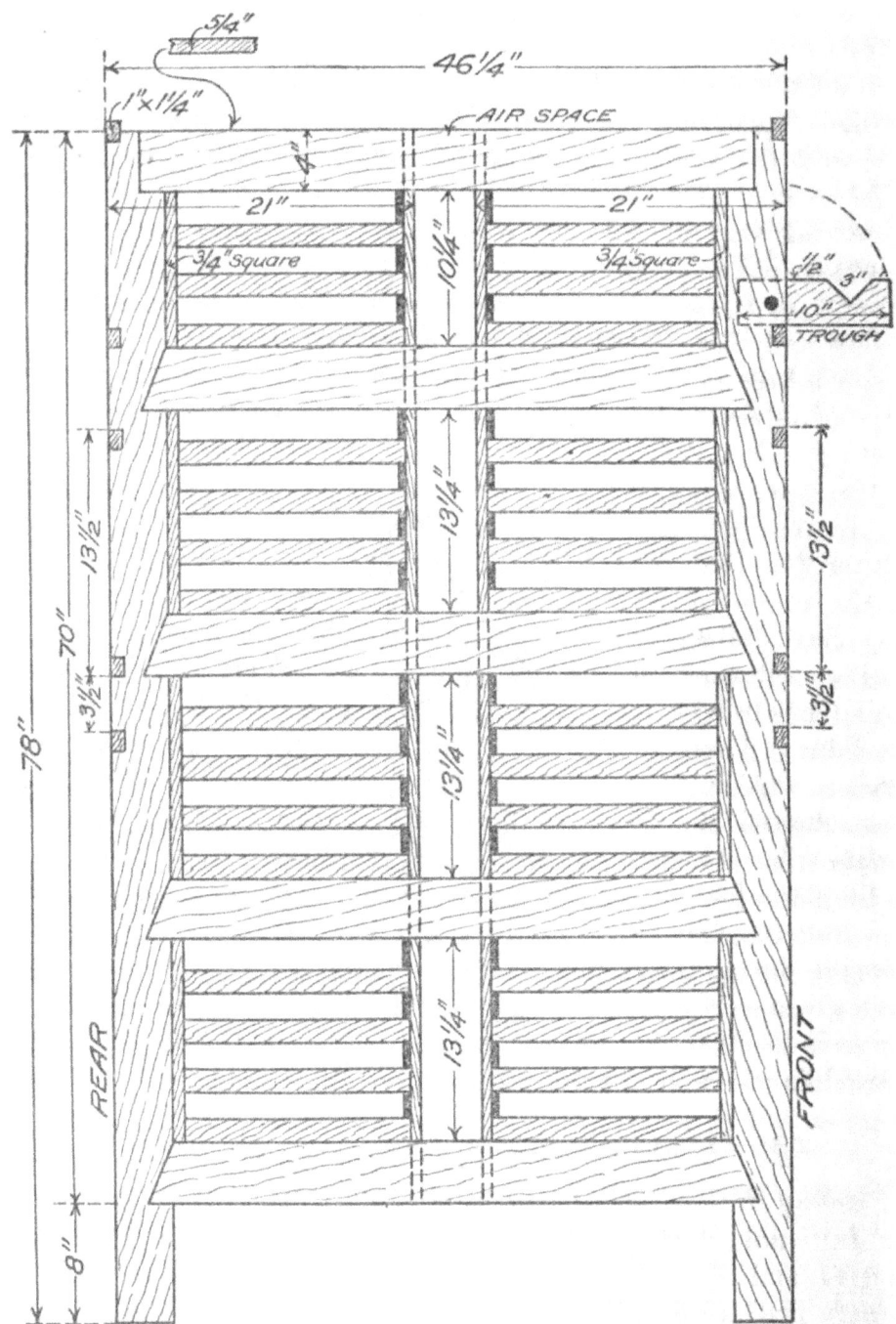

Fig. 1.—Stationary feeding battery, end view.

a lack of, or improper, ventilation. The following feeding stations were used for the experiments described in this bulletin; each one is designated by a number for convenience in reference;

STATION NO. 1.

This consisted of one section of a packing house known as the feeding station. The killing room where the poultry is dressed constitutes another section, and is situated on the side opposite the feeding station; the scales are located on the "dock" halfway between these two sections. This dock is a covered platform running the entire length of the packing house, and is devoted to the handling of the poultry and eggs. A spur track from the railroad runs parallel to the dock, so that the poultry and eggs can be unloaded directly from the car to the dock. The feeding station is 48 by 130 feet; 12 feet from the floor to the plate and 32 feet from the floor to the ridge, built on the "monitor" style. The walls are double, brick outside and plaster inside, and are sheltered by a projecting roof. The building contains 20 double-sash windows on both sides and 9 in both ends, with the sides of the monitor top entirely filled with sash

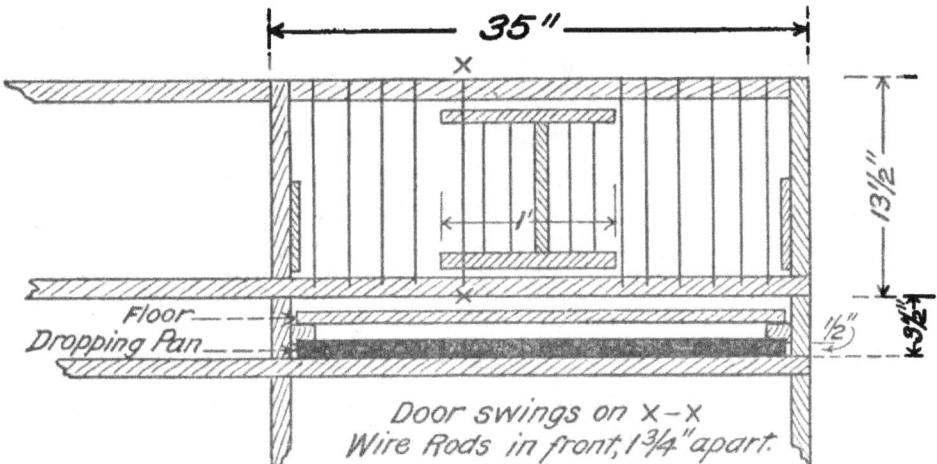

FIG. 2.—Front of stationary feeding battery.

and three double sash in both ends. A row of wooden shutters, 2½ by 2 feet, are placed on either side of the building just under the eaves. During hot weather the shutters are kept open and the windows all taken out and the openings covered with wire. The windows in the monitor top were originally hung at the top and opened in, but were changed about November 1 and hinged at the bottom. This change improved the ventilation in the house, as it tended to prevent the incoming air from falling directly on the birds in the batteries. About one half of one side of the house was built against another part of the packing house, and a dirt bank came up about 5 feet against one end, thus cutting off considerable ventilation during hot weather.

This house was equipped with stationary batteries, four tiers high, as illustrated in text figures 1 and 2. Eight chickens or 6 hens were placed in each division, and as each battery contained 56 pens, it would hold 448 chickens, but generally 450 birds were put in each battery. Later in the season, when the chickens brought in would weigh from 3 to 4 pounds, only 6 birds were placed in each compartment. The floors of each tier were 1-inch mesh wire, such as is ordinarily used for confining chickens, with a roosting board 3 inches wide, 2 feet 10 inches long and 1 inch thick, laid across the center of each division and set in between nails, so that it could be easily removed. The sides of the battery are wooden slats or laths, with a space of 1¾ inches between them, with wire rods set in the center of the furring in the front of the battery. The dropping pan is made of galvanized iron with the edge turned up in front and slides under each compartment, 3½ inches below the wire floor. There is a 4-inch air space which divides the battery into two parts, as shown in the accompanying cut. Feeding troughs made of cypress, 2⅝ inches on the side and 3¼ inches across the top, inside measurements, were hung along the front of each tier of coops, supported by wooden strips 10 by 3 inches, cut to fit the troughs, fastened with only one nail so that they can be pushed up into the battery when the troughs are removed. The center of the cut in this supporting strip is 2½ inches from the front of the battery, allowing about 1 inch between the front of the battery and the edge of the trough.

This house was arranged with a broad center aisle 8 feet wide, running lengthwise of the house, with side aisles between the batteries, which were 34 inches apart, and extended from the center aisle to the sides of the room. The side windows were at the end of the narrow aisles. This allowed the feeder to push the feed truck through the center aisle and one pail of feed would generally feed two tiers of birds, so that it only took four trips in each side aisle to feed the part of each battery which faces the aisle. This feeding station would accommodate 13,500 birds on feed at one time, with sufficient floor space for storing the grain and milk, mixing the feed, and storing the working equipment of the feeding station. The house was lighted by electric lights arranged so that the birds could see their feed on dark mornings and afternoons, and had four fans placed on top of the batteries to keep up a circulation of air in hot weather. These fans had to be cleaned frequently or they would become clogged with dust.

The feed was mixed in a portable feeding tank (illustrated in Pl. II, fig. 1), which is 2 feet wide, 6 feet 10 inches long on the top, 5 feet 6 inches long on the bottom, and 2 feet 1 inch deep. This tank is built of wood and contains a spout or valve at one end from which the feeding pails are filled. The truck ran on four wheels, a stationary pair at the end and two pivot wheels in front, so that it could be turned easily. The dropping pans were cleaned by scraping with a piece of galvanized iron 8 by 6 inches, one edge curved over to make a handle. A hook, made of bent wire, was used to pull out the droppings tray. The manure was scraped into a truck 21 inches wide, 4 feet long on the bottom, 4 feet 9 inches on the top, and 16 inches deep, with a removable slat across the top to rest the tray on while it was being cleaned. This truck contained one small pivot wheel on either end, with a pair of larger stationary wheels in the center. The portable crate or truck used for moving the birds is shown in figure 2 of Plate II. This is divided into two parts, with feed troughs in the center, and is covered with 2-inch mesh poultry wire. The crate is 5 feet 5 inches long, 5 feet 2 inches high, and 3 feet 2 inches wide, with doors hinged at the top, which swing in and are held open by a curved hook shaped like a letter U.

STATION NO. 2.

Feeding station No. 2, shown in Plate IV, is entirely separate from the other sections of the packing house. The building is 42 feet wide by 120 feet long, 14 feet to the eaves and about 15 feet to the top of the building proper, which has a monitor top 9 feet wide extending 5 feet 6 inches above it. Each side of the monitor top is divided into two tiers of shutters, 26 in each tier, hinged at the top, which swing out. Each side has 18 single-sash windows, set about 2 feet below the eaves, and 6 sash in either end. There are two tiers of shutters 2 feet 9 inches wide extending around three sides of the house, just above the floor, which are hung at the top and swing out. The east side—the building running north and south—has doors which slide up toward the eaves in place of the shutters. These shutters when tilted out shade the building during hot, sunny days. All of the shutters are made of $\frac{1}{2}$-inch matched lumber, and the sides of the house are built of ship lap $\frac{3}{4}$ inch thick and $5\frac{1}{2}$ inches wide.

Many of the birds in this house were sick, owing in some cases to their condition when they reached the packing house, but in most cases to drafts in the feeding station. The shutters had warped somewhat and did not fit tightly, so that the house was very drafty, as it was customary to keep part of the shutters open in the monitor top. This is a good house for summer feeding, as it can be thrown almost entirely open, so as to get all of the fresh air possible, but it is built too cheaply to make a good house for feeding in cold weather. This is quite apparent from the record of the deaths and of the gains obtained. If the shutters in the monitor top were hung at the bottom and swung in, the air would not fall directly on the birds below. During November heavy duck cloth was stretched partly across the building, about 2 feet above the top of the batteries, to cut off the drafts, which improved the house but did not allow sufficient ventilation. This style of house should be built so that the shutters on the side could be closed absolutely tight in cold weather and the draft cut off from the monitor top. Some ventilation is necessary even in cold weather to keep the air fresh.

This station was equipped originally with stationary batteries, but most of these had been replaced with portable feeding batteries, as shown in Plate III. This battery is divided into 8 coops, 4 tiers of 2 coops each, and holds 80 springers or 64 hens. It is 2 feet 7½ inches wide and 5 feet 9 inches high. The slats in the front are 1⅜ inches apart and each set of slats, which is 8¼ inches wide, is held in by buttons, so that it can be easily removed and a set of slats which are closer together or farther apart may be quickly inserted. As the sizes of the chicken's heads vary considerably during the season this changeable front is of value. The dropping pans are 1¾ inches below the floors, which are made of heavy, square-mesh wire, and have roost boards 2 inches wide by ¾ inch thick by 2 feet 6 inches long. The bottom of the first floor is 6 inches from the ground, and it is 15 inches from the wire floor to the top of each coop, making each tier, including the dropping pans, 16¾ inches deep. The battery rolls on four wheels, two double-pivot wheels in front and two wheels connected by a bar in the rear. The sliding doors on the sides are fitted with hooks which fasten into eyes on the battery. The whole battery is made of furring, 1⅛ by ⅞ inches, covered with 2-inch mesh wire and laths. The feeding troughs are 3½ inches across the top, inside measurement, and 3 inches from the top edge to the bottom, outside measurement. These troughs are held in place with bent wires which are flexible so that they give if the troughs hit any obstacle, thus preventing breakage. A wire partition divides the battery into two equal parts. A similar battery is used for feeding turkeys except that it contains three tiers instead of four and the slats in front are 2⅝ inches apart. This battery is illustrated in Plate III, figure 3.

These coops were arranged in long rows running lengthwise of the house, spaced about 4 inches apart in the rows, with the rows 3½ feet wide. This arrangement was changed to suit the varying conditions. The arrangement of the batteries in long rows tended to waste labor, as the attendant would feed down a row till he emptied his pail and then return to the end for another pail of feed, thus making many trips with an empty pail. This could be overcome by arranging the batteries differently, or by having feed at both ends of the rows. The batteries of birds to go on feed were pushed in at one door and rolled out at the other end, which made it necessary to keep moving the batteries in each line down toward the end of the house. This frequent moving is detrimental to good results in feeding as it keeps the birds restless.

DESCRIPTION OF FEEDING STATIONS.

Various styles of pails were used in feeding, two good types of which are shown in Plate III, figure 4. These pails hold about 14 quarts.

The best style for a feeding pail depends somewhat on the thickness of the feed and the style of the battery in use; the styles in Plate III, figure 4, were adapted to the portable feeding batteries and were used at stations 2, 3, and 4. The feeding pail used at station 1 had a blunt snout 4 inches wide, which projected out 4 inches at the top and tapered to nothing at the bottom of the pail. This held 18 quarts and had to be handled carefully or the feed would come out too quickly. It would not be adapted to use with portable batteries, as much feed would be wasted between the troughs. The spout of the shorter bucket, illustrated in Plate III, figure 4, is too small for thick feed and makes slow feeding, but the other style of bucket is very satisfactory. This bucket is $12\frac{1}{2}$ inches deep and 9 inches in diameter. The small end of the spout is $1\frac{5}{8}$ inches in diameter. Some pails have a handle on the back near the bottom of the pail.

The manure truck shown in figure 3 of Plate II was used at stations 2, 3, and 4. All of the stations were equipped with some kind of spraying machine, fitted with small air-pressure tanks, which were operated by hand power. A small amount of carbolic acid or some similar disinfectant was generally mixed with the whitewash.

STATION NO. 3.

The best results were secured in feeding station No. 3, which is shown in Plate IV, figure 2. This is a lean-to, built against one side of the packing house, and is a very inexpensive building. The floor is a single layer of unmatched boards, and the sides are of standard duck cloth, 10-ounce weight. The curtains overlapped each other about a foot and were arranged so that they could be easily rolled up and down. Three sides of the house could thus be entirely opened to allow perfect ventilation. The floor should either be double or else made of matched lumber, as the single unmatched boarding allows too many cracks. This shed was protected on all sides by other buildings, so that it did not get the full force of the wind. The birds were very free from colds and disease in this building, and it appeared to be an ideal place for summer and early fall feeding.

The curtain idea could be adapted to other feeding stations with good results, if care is taken to see that the house is made free from drafts. This house did not contain any windows. The floor was 45 by 93 feet, with the roof 9 feet from the floor on the lower side

30 FATTENING POULTRY.

and 13 feet from the floor where it was attached to the packing house. The house was equipped with portable feeding batteries arranged in short rows with a center aisle 5 feet wide running the long way of the building and side aisles 4 feet 6 inches in width. These batteries were changed to suit conditions. This station was separate from the other parts of the packing house, so that the birds were disturbed only at feeding and cleaning times. The feed was mixed in a long stationary mixing tank and carried to the feeding room in pails on trucks.

STATION NO. 4.

The entire second floor of the packing house was built for a feeding room in station No. 4. This is a new station and contains some very good ideas. It is illustrated in Plate IV, figure 3. This feeding room is built with a wing; the main part of the house is 48 by 140 feet with the wing 48 by 48 feet. The house is equipped with portable batteries similar to those at stations 2 and 3, and arranged as at station 3, with a center aisle 4 feet wide and side aisles 3 feet 10 inches wide, but subject to change according to conditions. This station was equipped with a mixing machine (described on page 20) and had an elevator large enough to hold two portable feeding batteries, which connected with the killing and weighing room on the first floor. The first floor was divided into a killing or picking room, a packing and small cold-storage room, an office, and general space for weighing, storing equipment, etc. Both floors of this building were made of cement.

The feeding room has an almost flat roof about 15 feet from the floor, and all sides of the room except the west contain two tiers of shutters each 4 feet 3 inches high and one tier of windows 3 feet high, so that the room can be thrown almost entirely open. The west side contains one tier of shutters and one of windows. The shutters are hung at the top and swing out from the bottom, while the windows, which are glazed, are hung in the center. The shutters are made of narrow strips $\frac{7}{8}$ inch thick, laid diagonally, and the whole room is well built. This station was not occupied until November, but it is apparently a very good type of feeding station and contains many excellent features. The ventilation in warm weather ought to be ideal and yet the building can be shut absolutely tight in cold weather if desired.

The birds were moved into this station from station No. 5, shown in figure 4 of Plate IV.

STATION NO. 5.

Station No. 5 is a combination of feeding station and creamery, the second floor being the feeding station. This house was equipped with long stationary batteries, built somewhat similar to those described in station No. 1, except that the batteries were only three tiers high. The wires in the front of the coops were 1¾ inches apart. The center aisles were 3 feet 9 inches and the side aisles 2 feet 6 inches wide.

DETAILS OF THE FEEDING EXPERIMENTS.

The records which are shown in detail in Tables I and II of the Appendix represent results obtained at different feeding stations in the Middle West during the season of 1910. The tests cover a large number of birds, and the conclusion derived from the averages should be of considerable value, and should largely eliminate the error which is almost certain to be made in drawing conclusions from experiments dealing with small lots. The variation found to occur within a lot kept under similar conditions during the feeding test clearly indicates how great the error may be if conclusions are drawn from results secured in dealing with small numbers. Undoubtedly, however, the conditions existing at each feeding station, outside of the rations used, had considerable influence on the gains secured.

Experiments A and B give the results of each lot in detail, while experiments C and D only give the number of birds, the length of the feeding periods, and the gains in terms of percentages. The total weight of the birds before they were fed in these records is 793,359 pounds, 303,222 pounds of which are included in the detailed experiments A and B.

In experiment A the number of dead is the difference between the "Number in" and the "Number out," but in experiment B a large number of crippled birds were removed and dressed, as is more fully explained hereafter, and their weight is credited to the particular lot in the column headed "Gain" and in "Per cent of gain," but is not included in the "Weight out" column. The column headed "Average grain daily per head" represents the average daily feed (not including milk) of each lot for its feeding period. This factor is obtained from the daily feeding reports which give the total daily consumption and the number of all the birds on feed for each day, and is based on the assumption that all the birds eat the same amount of feed. The total feed for each lot is derived from this factor. This method of obtaining the total feed is not absolutely correct for each lot, but it is the only method which is practicable under commercial conditions which deal with large numbers, and the possible error would not affect the averages, but would help to explain the differences in the gain of certain lots which probably averaged to eat different amounts of feed per bird.

32 FATTENING POULTRY.

The following prices were used in figuring the feed cost of grain:

Corn meal	$1.35	per 100 pounds.
Low-grade wheat flour	1.35	per 100 pounds.
Oat flour	2.20	per 100 pounds.
Tallow	.08	per pound.
Buttermilk	.015	per gallon.
Condensed buttermilk	.75	per 100 pounds.

The average daily labor cost in experiment A was $7.29 per 10,000 birds on feed, which constant was used throughout the feeding season. The labor cost per 10,000 head varied considerably during the season in experiment B and the following constants were used: $9.44 per 10,000 head for lots 1 through 19; $9.53 for lots 20 through 41; $12.06 for lots 42 through 63, and $15.63 for lots 64 through 83. The cost of 100 pounds of grain, plus the cost of the milk used in feeding that amount of grain, varied as follows in experiment A: $1.93 per 100 pounds of grain for lots 1 to 20; $2.06 per 100 pounds for lots 21 to 37, and $1.95 per 100 pounds for lots 38 to 63. The cost in experiment B was $2.31 per 100 pounds of grain for lots 1 through 19; $2.30 for lots 20 through 41; $2.43 for lots 42 through 63; and $2.44 for lots 64 through 83. The grain in experiment A cost $1.35 per 100 pounds throughout the feeding period, while the cost of the milk used with 100 pounds of grain varied from $0.585 to $0.709. The cost of 100 pounds of grain in experiment B varied during the feeding season from $1.94 to $2.17, and the cost of the milk used with 100 pounds of grain varied from $0.275 to $0.352. Condensed buttermilk was fed in experiment A in a much thicker state than ordinary buttermilk, which explains the increased cost of the milk in experiment A over that in experiment B. The average cost of 100 pounds of grain in experiment A for the entire season was $1.35; and in experiment B, $2.06; while the average cost of the grain and the milk per 100 pounds of grain was $1.98 in experiment A and $2.37 in experiment B.

EXPERIMENT A.

The feeding was conducted at station No. 3. All of the lots in this experiment were fed alike except that the length of the period varied from 6 to 10 days. These birds were only fed for this short period because the existing conditions were such that it was not convenient to keep them on feed for a longer time. The gains secured indicate that under good conditions a large gain can be made in a short time.

Stock.—The lots were composed of stock of mixed origin put into the feeding batteries just as they came in from the small live-poultry buyers, without sorting. Each lot contained birds of various weights, but, as the records show, the average size or weight increased as the

DETAILS OF FEEDING EXPERIMENTS.

feeding season advanced, although there were some birds of broiler size in practically all of the lots. The Barred Rocks were the most popular breed; records kept at various intervals showed that about 42 per cent were Plymouth Rocks and 17 per cent were Leghorns. None of the other breeds were represented by large numbers of birds, but numerically they were present in the following order: Wyandottes, Rhode Island Reds, Orpingtons, Minorcas, and Langshans, constituting altogether not over 10 per cent of the total number of birds on feed. Thus 31 per cent of the birds were of mixed breeding. These figures only represent percentages in the rough, as the birds were classed as Plymouth Rocks, Leghorns, etc., if they had the most prominent characteristics of these breeds; many of them were probably grade stock.

These figures, moreover, represent averages of large numbers, and do not necessarily represent the actual composition of any specific lot.

The average quality of the stock in experiment A was good, apparently slightly better than in experiment B and considerably above that in experiments C and D. This means that the majority of the birds were slightly better fleshed and that there were fewer sick birds than in the other experiments. The difference in the health of the stock was hardly apparent when the birds were received at the feeding stations, but showed up noticeably during the feeding period, although the housing and management undoubtedly affected the health at the different stations, thus introducing a new factor, which must be considered in making comparisons.

Handling.—Most of the birds in experiment A were shipped by express to the feeding station, not over one-third arriving by freight. In general the distance shipped was short. The birds shipped in by express were weighed and put into the portable feeding batteries, previously described, shortly after reaching the feeding station. Those shipped by freight were handled in crates in stock cars, and generally stood several hours at the packing house before they were unloaded and put into the feeding batteries. The use of a portable feeding battery eliminates labor to a considerable extent, and involves less handling of the birds, both when they go into the feeder and when they come out. The birds undoubtedly get into the feeder in better condition, and there is less chance of breaking the wings after they leave the battery. After fattening, the batteries of birds were taken directly into the killing room, where they were taken out by the pickers as killed.

Feed.—Ration No. 1 was fed to all of these birds, the grain and buttermilk being mixed with a rake in a large tank. The buttermilk used was condensed and was diluted with about 2 parts of water to 1 of buttermilk. A small amount of whey and considerable skim milk was fed during the season, replacing some of the water, and, at times, a part of the buttermilk when the supply ran short, as happened during parts of October and November. Granulated or shredded curd was added to the ration several times during September and October, and the birds appeared to relish it very much. No grit was provided for the birds in this experiment.

The combination of condensed buttermilk and skim milk with the grain made a very thick feed, which was eaten very eagerly by the chickens, but it was necessary to give a thinner feed once each day, which was done by increasing the proportion of water. The chickens were not so eager for the feed when whey was used to replace much of the buttermilk, which often happened during October and November.

Method of feeding.—The chickens were fed three times daily, generally receiving the thinner feed at noon. Wheat flour has to be added very gradually in making the mixture, otherwise it will lump, and the person who mixed the feed often found it necessary to knead it as the housewife kneads dough in making bread. The feed was run into pails, which were carried into and through the feeding room on trucks. Generally two or three persons handled the feed, one filling and distributing the pails while the others did the feeding. One man fed part of the birds regularly three times a day, but he had different assistants at various times during the feeding season, no other help being kept specially for this work. The results, as shown in the record, indicate how well this system of feeding is adapted to conditions where the help is not especially experienced in feeding chickens. A moderate amount of feed was poured into each trough, and by the time the feeders had fed all the birds once those fed first had eaten up their supply and were looking for more. The feeder then gave a second feed, which was a light one, only to those birds that had cleaned up their first feed. Generally 20 to 30 minutes intervened between the first and second feed, according to the length of time it took the feeders to go through all of the aisles.

Sometimes the feeder would go around the third time, but not generally. This method of feeding appears to stimulate the appetites of the birds so that they consume a large amount of feed; and it does not require as skilled a feeder as is necessary when the birds are fed only once. There is also less chance of feed being left over in the troughs. The feeder went through the batteries about one hour after feeding, removing any feed which was not cleaned up by that time. However, if care is exercised with this method of feeding there should be no feed left over to be cleaned up. When the birds clean the feed up quickly the troughs are left in good condition, but if any feed is left over the feeder is apt to leave a little in the trough when he scrapes out the surplus, and this is likely to ferment or become sour. Therefore, the aim should be to feed only what the chickens will eat up clean. When the feeding is managed properly the feeding troughs appear as though they had been washed after the birds are through eating. The feeding troughs are not washed under ordinary commercial conditions except in special cases. When the batteries are sprayed the troughs receive more or less whitewash, but no special care is taken to clean them, as they are not dirty ordinarily.

The batteries were arranged to suit the convenience of the operator and to conform to the size of the room. The aim was to give the birds the best possible conditions without making the labor cost excessive. During the first part of the feeding season most of the birds were fed in a general purpose room, used for weighing all of the live and dressed poultry, for storing and mixing the feed, for batteries of hens held only one or two days before killing, and for candling eggs. In other words, there was something going on in this room practically all of the time, still the birds made good gains. During this time a few of the batteries were kept on the open dock where there happened to be a little available space. A new feeding shed was built, into which the batteries were moved on October 22, where the birds were isolated and so were disturbed only at the feeding times. There was no marked change in the gains due to this change in the feeding rooms, which appears to indicate that while it may be better to have the birds in a quiet, secluded room, this point is not as essential as is generally supposed. Most articles on fattening recommend that the feeding station be kept dark except at feeding times, but many of the large poultry packers pay no attention to this matter, and results appear to indicate that it has no important bearing on the fattening question. It is impossible to draw definite conclusions about such questions from the differences in the gains, as there are so many other factors subject to constant variation which affect the results.

The batteries were arranged in lines with an aisle running through the center of the building from end to end, as described under station No. 3 (p. 30). Batteries put on feed at the same time were kept together for convenience in handling. The batteries were placed several inches apart to allow a good circulation of air, the distance apart depending both on the temperature and on the amount of available floor space. One great advantage of the portable batteries is that they may be spread around to suit varying conditions. As birds often suffer from the heat during excessively hot weather, and many deaths occur, this point affects gains quite materially. The influence of the plan of the house on the gains is discussed more in detail under "Feeding stations." These birds were housed in station No. 3 after October 27. The batteries should also be arranged for convenience in feeding, so that a pail will contain feed enough to go up and down each side aisle and bring the feeder back to his source of supply, thus avoiding a waste of time in carrying an empty pail. The method employed in this house was to have a wide center aisle with branch aisles five or six batteries deep.

Table I of the Appendix shows the results of the feeding in experiment A in detail; a summary is given in Table 3, below.

TABLE 3.—*Summary of feeding experiment A, arranged according to length of feeding period.*

Number of head.	Days fed.	Average weight.	Per cent of gain.			Grain per pound of gain.		
			High.	Low.	Average.	High.	Low.	Average.
		Pounds.	Per ct.	Per ct.	Per ct.	Pounds.	Pounds.	Pounds.
2,068	10	2.51	23.5	11.5	18.5	5.32	3.01	4.04
10,360	9	2.40	26.1	11.2	19.4	5.10	2.55	3.52
11,878	8	2.55	27.1	10.9	17.2	4.40	2.17	3.37
15,731	7	2.39	29.6	11.4	19.2	4.55	1.92	2.68
3,907	6	2.18	18.6	8.2	13.1	5.35	2.14	3.66
43,944	....	2.42	29.6	8.2	18.1	5.35	1.92	3.26

Number of head.	Cost of labor per pound of gain.			Cost of feed per pound of gain.			Total cost per pound of gain.		
	High.	Low.	Average.	High.	Low.	Average.	High.	Low.	Average.
	Cents.	Cents.	Cents.	Cents.	Cents.	Cents.	Cents.	Cents.	Cents.
2,068	1.95	1.43	1.67	10.37	5.81	7.84	12.32	7.24	9.51
10,360	2.09	.99	1.51	9.95	4.97	6.88	11.77	5.96	8.39
11,878	1.86	.92	1.39	8.58	4.23	6.64	10.12	5.15	8.03
15,731	2.31	.88	1.17	8.78	3.71	5.42	11.09	4.61	6.59
3,907	2.81	.98	1.73	10.39	4.17	7.28	13.14	5.15	9.01
43,944	2.81	.88	1.40	10.37	3.71	6.45	13.14	4.61	7.85

The average daily amount of grain consumed per head in the above experiment was as follows: High, 0.2593 pound; low, 0.1132 pound; average, 0.1766 pound. The total weight of the birds was 130,430 pounds.

SUMMARY OF EXPERIMENT A.

The cheapest gains and the lowest average cost of gain in this experiment was made by the lots fed 7 and 8 days. The difference in cost of gain between the 6-day lots and the 7-day lots is quite marked, while the 10-day lots show a considerably increased cost over those fed 9 days. As there was a much smaller number of birds in the 6 and 10 day lots there is a considerable possibility of error in drawing conclusions from a comparison between them and the lots fed 7, 8, and 9 days. Other conditions being equal, the lighter birds make greater gains than the heavier birds, which would partly account for the very slight difference in gain between the 8 and 9 day lots, as the average weight of the 8-day lots is much greater than the weight of the 9-day lots. This would also tend to cause the increase of the cost of gain in the 8-day lots as compared with the 7-day lots. This table shows a marked advantage in 7-day feeding over either a shorter or a longer period in producing cheap gains, but the fact that the quality of the flesh and the appearance of the bird improves with the length of the feeding period, as previously shown, should be considered in determining the best length of time to feed the birds.

EXPERIMENT B.

The lots in experiment B were fed from 6 to 15 days, the exact time depending on the season, the kind of birds, and the methods of the feeder. The feeding was carried on at station No. 1. The relative gains secured in the shorter feeding periods show distinctly that the daily gains secured in the first 6 or 7 days of the feeding period are greater than those secured on the succeeding days. The gains secured on the 7 or 8 day lots, as against the 13 or 14 day lots, show that from the standpoint of cost of gain alone the profit is much greater in the lots fed for the shorter periods. The market, however, affects to a considerable extent the best length for the feeding period, as the class into which the birds are placed when dressed depends to some extent on the length of the feeding period. Thus in order to turn out a lot as broilers it is only necessary to feed for 7 or 8 days, while if kept on feed for 13 or 14 days the same lot would be put into a different class, for which there might not be as good a demand. But allowing for all these influencing factors, the results indicate the advantages of short feeding periods with the method and under the conditions of this experiment. It is true, however, that the birds become more uniform and show the effects of milk feeding more plainly when kept on feed for the longer periods.

Stock.—This stock was very similar to that described under experiment A, but not of quite as good quality. The difference is apparent in the larger percentage of dead birds, but it is still more marked in the number of so-called "cripples." The cripples are birds off feed, which are taken from the lot and dressed during the feeding period. Their weight is credited to the dressed weight of each lot in securing both the gain in weight and the per cent gain, and it represents the difference between the gain as shown by the "weight-out" column and the number of pounds gain. The number of dead was greater in this experiment than in experiment A. In experiment A no cripples or sick birds were removed. The dead birds were picked up each day in both cases, but in experiment B special care was taken to remove all cripples as well as sick and dead birds. When sick birds occur in feeding lots this part of the work is very essential, as sickness among birds may spread rapidly where such a large number are kept together. As a general rule the amount of sickness under ordinary feeding-house conditions is small. The gains secured during the last half of September and all of October show very markedly the effect of sickness among the birds. During this period and part of November it is hard to secure good results in fattening under ordinary commercial conditions, as the chickens are apt to have colds and may develop other troubles. But the conditions under which the birds have been reared, as well as those at the feeding station, control this question.

In this experiment the lots were sorted, beginning about the middle of September, into roasters, broilers, and springs. The roasters represented the heavier chickens, the broilers the light-weight birds, with the class between termed springs. Lots which were not sorted are classed as springs in all of the records, so that this class includes many different weights of birds. While there is a great variation in the percentage of gains, the average results show that light-weight birds gain a much larger per cent than the heavier birds in the same length of time on feed; and in most cases the broilers have gained as large a per cent in 7 or 8 days as the roasters gained in 13 or 14 days. When the birds at this station were sorted, the roasters were generally fed for the longer period and the broilers for the shorter period. One reason for feeding the broilers the shorter period was because the feeder desired to turn out as many broilers as possible and still have the birds in good condition.

The breeds represented in these lots are the same as those in experiment A, and the relative proportions are quite similar. The records show that 16 per cent were Leghorns or birds belonging to the Mediterranean class, but no record was kept of the proportion of the other breeds present.

Handling.—Over three-fourths of the birds were shipped to this station in coops in live-stock cars, only a few coming in by express. Most of the birds were shipped a longer distance than those in experiment A, and many were shipped by freight although coming only a short distance. A car used entirely for poultry and eggs was switched to this station nearly every day, containing most of the eggs and poultry shipped along a branch railroad line. No live-poultry cars were received at either station No. 1 or No. 2 during the period covered by these records.

The chickens were weighed on the dock, put into a transfer battery, and carried into the feeding station, where they were placed in the stationary batteries. At the end of the feeding period the chickens were again put into the transfer batteries, weighed, and rolled into the killing room. This method of handling involved considerable labor and more handling than the method of transferring the birds described under experiment A. It was, however, preferred by the feeder of this station, who claimed that it kept the station cleaner and freer from insect pests than the other method. No trouble was apparent in station No. 3 due to the causes mentioned. The question seems to be largely one of personal preference, but the writer prefers the portable feeding battery. Birds in the stationary batteries can be fed more easily and quickly and with less spilling of feed than those in the portable batteries, as the trough in the first case is continuous, while in the latter the line is broken every 2 or 3 feet. It is also easier to scrape back feed from the longer troughs used in the stationary batteries. The stationary battery is the older and more common method of handling birds.

Feed.—Ration No. 2 was used in feeding all the lots in this experiment. This is a very good ration, and is quite similar to the one fed to the lots in experiment A except that oat flour is used in place of low-grade wheat flour and a small amount of tallow is added. The grains were mixed with ordinary buttermilk rather than condensed milk. A comparison of the results secured in these two sets of records appears to indicate that while oat flour produced as good or even better gains than wheat flour, the wheat flour gave more economical gains, and therefore was the more practical feed to use, considering the relative cost of the grains. Tallow appears to be almost too expensive to feed economically, considering the possible detrimental effect which it may have on the quality of the dressed poultry. A comparison of the results with those which follow makes one wonder how important a part the feed has in effecting the gains, as in many of those cases the cost of the gain is greater than in the present experiment, although the ration and general method of feeding were the same. Still the conditions of the feeding in experiments A and B were apparently about equal, and the conclusions drawn in a comparison of the relative value of grains should not be much out of the way if allowance is made for some differences noted elsewhere in this bulletin.

The feed was mixed by hand with a rake in portable feeding trucks (see Pl. II, fig. 1). The proportion of milk in the feed varied considerably from day to day, closely following the changes in the weather, a larger quantity being fed when the weather was hot or excessively dry. Ordinarily the amount of milk varied from 60 to 70 per cent, although in a few instances it went either much lower or much higher than these figures. The average percentages fed during certain periods were as follows: July 28 to September 9, 65.3 per cent; September 10 to October 7, 67.5 per cent; October 8 to November 6, 64 per cent; November 7 to December 5, 62.3 per cent.

The feed ordinarily appeared considerably thinner at this station than was the case in experiment A. This was due largely to the fact that low-grade wheat flour mixed with milk makes a thicker, stickier mass than an equal weight of oat flour mixed with the same amount of milk; there was, moreover, a somewhat larger percentage of milk in the ration used at this station. The birds seemed to like the thicker feed, but this principle could easily be carried to an extreme, and, as discussed previously, at least one feed daily should be relatively thin. Melted tallow was added to the mixture of grain and milk. Grit was given twice a week between regular feeding times.

Feeding.—These birds were fed twice daily, at 6.30 in the morning and between 2.30 and 4 or 5 o'clock in the afternoon, depending on the weather and on the appetite of the birds, which probably was more or less influenced by the weather. If the birds became active and restless early in the afternoon, indicating that they were hungry, the feeding was begun earlier than when they showed no desire for food until late in the day. The birds were fed only once at each feeding time, receiving a liberal feed. The feed was taken away if not cleaned up two hours after feeding in the morning but was left at night, and this feed was always cleaned up by the chickens before feeding time the next day. This method requires a more experienced feeder than the method of feeding three times daily and refeeding at each feed, as more judgment is required in regulating the amount to feed, but it gives good results if done carefully. Most beginners and inexperienced persons would get the best results by using the method described under experiment A, and on the whole it seems to be the preferable way. However, where the birds are fed only twice they are quieter during the day than if fed three times. The conditions in this feeding station tended to keep the birds quiet, as there was no confusion outside of the regular work of feeding and cleaning. No special effort was made to keep the birds in the dark. This station, like all the others, was equipped with electric lights, which were used whenever the room was dark at feeding time.

It was observed that the birds in the lower tiers invariably ate better than those in the upper tiers, probably owing to the fact that it is cooler in the lower tiers, as a large amount of heat rises from the chickens. The fact that this difference is most marked during hot weather appears to confirm this theory. The lower tiers are darker and the birds more secluded, which may also aid somewhat in fattening.

TABLE 4.—*Summary of experiment B, arranged according to length of feeding period.*

Number of head.	Days fed.	Average weight.	Per cent gain.			Grain per pound of gain.		
			High.	Low.	Average.	High.	Low.	Average.
		Pounds.	Per cent.	Per cent.	Per cent.	Pounds.	Pounds.	Pounds.
892	15	1.90	38.7	38.7	38.7	2.58	2.58	2.58
6,720	14	2.18	57.8	17.8	36.8	4.00	1.86	2.80
8,464	13	2.54	31.0	15.6	23.2	4.20	2.27	3.16
1,657	12	3.22	13.3	11.5	12.4	4.74	3.77	4.26
644	11	3.98	10.6	6.7	8.7	6.49	3.65	5.07
7,836	10	3.24	26.7	7.7	13.6	5.84	2.12	3.71
7,368	9	3.18	36.2	4.0	14.7	8.19	1.95	3.66
12,199	8	2.81	31.4	5.2	17.0	8.45	1.82	3.16
14,841	7	2.69	40.5	5.3	16.4	5.05	1.29	2.96
1,085	6	3.08	16.3	6.8	11.6	3.31	3.61	2.96
61,706		2.82	57.8	4.0	18.7	8.45	1.29	3.26
[1] 7,753		2.30	36.2	5.4	17.7	8.45	1.61	3.28
[2] 18,864		3.61	29.3	4.0	11.3	8.19	2.27	4.18
	6–10				14.7			3.49

Number of head.	Cost of labor per pound of gain.			Cost of feed per pound of gain.			Total cost per pound of gain.		
	High.	Low.	Average.	High.	Low.	Average.	High.	Low.	Average.
	Cents.	Cents.	Cents.	Cents.	Cents.	Cents.	Cents.	Cents.	Cents.
892	1.91	1.91	1.91	5.96	5.96	5.96	7.87	7.87	7.87
6,720	2.57	1.38	1.96	9.20	4.30	6.47	11.77	5.68	8.43
8,464	3.33	1.47	2.26	9.70	5.22	7.28	13.03	6.69	9.54
1,657	3.00	2.58	2.79	10.90	8.67	9.79	13.90	11.25	12.58
644	4.60	3.13	3.87	15.77	8.87	12.32	20.37	12.00	16.19
7,836	4.64	1.41	2.88	14.19	4.88	8.92	18.33	6.29	11.80
7,368	7.30	1.28	3.20	19.90	4.49	8.77	27.20	5.77	11.97
12,199	5.73	1.32	2.60	19.44	4.20	7.54	25.17	5.52	10.14
14,841	5.06	1.14	2.44	12.32	3.15	7.07	17.38	4.35	9.51
1,085	3.28	1.63	2.46	8.08	6.00	7.04	11.36	7.63	9.50
61,706	5.63	1.14	2.59	19.90	3.15	7.74	27.20	4.35	10.33
[1] 7,753	5.73	1.47	2.47	19.44	3.93	7.68	25.17	5.60	10.15
[2] 18,864	7.30	1.47	3.41	19.90	5.22	10.01	27.20	6.69	13.42
			2.72			7.87			10.59

[1] Broilers. [2] Roasters.

The average daily amount of grain per head consumed in the above experiment was as follows: High, 0.2007 pound; low, 0.0758 pound; average, 0.1449 pound. The total weight of the birds was 172,792 pounds.

SUMMARY OF EXPERIMENT B.

The cheapest gain of any individual lot in this experiment was made by a 7-day lot, but the lowest average cost of gain was made by the 14 and 15 day lots. The table shows clearly that the cost of gain is in direct proportion to the average weight of the birds, which partly accounts for the cheapest gains in lots 14 and 15. The comparison of the 15-day lot with the 14-day lot is hardly fair, as only a small number of birds were fed 15 days, which was also true of the 6 and 11 day lots. The highest gain was made by a 14-day lot, which shows that very good gains can be made in the longer feeding periods. A study of the table as a whole would indicate that the results secured in 13 or 14 day feeding periods were as profitable as those obtained in the 7 and 8 day periods, allowing for the modifying conditions already noted. A comparison of the roasters and broilers shows that the broilers make the cheaper gains, which emphasizes the conclusion already drawn that the best gains are made by light birds, as all of the broilers were fed for a relatively short period, while about half of the roasters were fed for 13 or 14 days and the rest from 7 to 9 days.

COMPARISON OF EXPERIMENTS A AND B.

Comparing experiment A and experiment B it is seen that the variations in the latter are much greater than in experiment A, which shows that birds vary greatly in their ability to put on flesh, and that this variation is most marked in the longer feeding periods. It took the same average amount of grain to produce a pound of gain in both experiments, but the average total cost of producing a pound of flesh is 2.48 cents greater in experiment B than in experiment A, while the average food cost is 1.29 cents greater. Comparing lots fed from 6 up to 10 days in both experiments it is found that the average gain was 3.4 per cent greater, the grain per pound gain 0.23 pound less, the cost of labor 1.32 cents less, the cost of feed 1.42 cents less, and the total cost 2.74 cents less in experiment A than in experiment B. Therefore the ration in experiment A was a cheaper feeding ration than that used in experiment B, or, in other words, low-grade wheat flour produced cheaper gains in fattening chickens than oat flour. However, the increased amount of milk and the different methods of management undoubtedly affected the results to some extent. The cost of labor is a very important item and materially affects the cost of gain. The method of feeding in experiment A appears to be better than in experiment B, showing that feeding three times daily is better than feeding twice daily. All of these factors are closely related, which makes it impossible to determine the relative importance of each.

EXPERIMENT C.

The quality and condition of the stock in this experiment were fair, although a little lower in quality than the stock in experiments A and B. The housing and feeding conditions were slightly unfavorable during most of the period, as the birds were housed in a rented feeding station, previously described as station No. 5, until about the middle of November; after that the feeding was carried on at station No. 4. As the company intended to move their birds as soon as the new plant could be erected, no special effort was made to improve conditions in the rented feeding station; consequently much of the work was done at a disadvantage, which undoubtedly affected the results to some extent.

The lots in this experiment were fed ration No. 1, except that the proportion of corn meal and flour was varied throughout the feeding season, and that during the latter part of the period covered by this record the amount of corn meal was gradually increased about 10 per cent. A small proportion of shorts, varying from 6 to 12 per cent, with an average of 10 per cent, was added to the ration throughout the whole period. This ration was mixed with ordinary buttermilk obtained directly from a creamery, the proportion of milk in the ration varying from 58 to 65 per cent, with an average of 62 per cent for the period. These changes do not materially affect the cost of the ration.

The length of the feeding period varied from 5 to 14 days. The gains show considerable variation in the different lots, but taken as a whole the conclusions drawn from the previous records apply to the lots in this experiment. It should be stated, however, that the conditions at this station and at the station where experiment D was conducted were not studied as closely as those at the two stations represented by experiments A and B. The method of feeding was similar to that used in experiment A, except that the birds were given a very light feed at noon and not fed a second time at that meal. A tendency to leave feed in the troughs before the birds all of the time was observed. The feed was mixed by hand until the chickens were moved into the new building, where two feed mixers were installed. This feeding station (No. 4), previously described, represents the result of study and extensive experience by the manager, and contains many excellent features.

TABLE 5.—*Showing per cent gains made by chickens in experiment C, by months and by length of feeding period in each month.*

Month.	Number of head.	Days fed.	Per cent gain.		
			High.	Low.	Average.
			Per cent.	Per cent.	Per cent.
July	5,622	14	43	20	27.3
August	18,209	14	44	9	25.9
September	5,736	14	41	12	30.4
Do	1,374	13	20	10	15.0
Do	2,284	12	22	12	17.7
Do	8,477	8	28	7	15.8
Do	6,150	7	19	7	19.7
Do	6,069	6	20	2	11.9
Do	3,879	5	22	0	13.0
October	7,995	14	55	22	36.6
Do	1,765	13	41	26	32.0
Do	2,746	12	34	25	30.6
Do	2,139	11	39	31	34.0
Do	1,887	8	12	10	11.0
Do	8,854	7	16	3	9.7
Do	7,208	6	21	2	12.5
Do	3,160	5	21	1	10.5
November	1,616	14	25	11	17.8
Do	5,041	13	37	15	27.5
Do	1,175	12	27	24	25.3
Do	1,265	8	16	13	14.5
Do	6,566	7	17	8	12.6
Do	3,980	6	14	5	9.7

TABLE 6.—*Summary of experiment C, arranged according to length of feeding period.*

Number of head.	Days fed.	Per cent gain.		
		High.	Low.	Average.
		Per cent.	Per cent.	Per cent.
39,178	14	55	9	27.6
8,180	13	41	10	24.8
6,205	12	34	12	24.5
2,139	11	39	21	34.0
11,629	8	28	7	13.8
21,570	7	19	3	14.0
17,277	6	21	2	11.4
7,039	5	22	0	11.8
113,217	....	55	....	20.2

The average quantity of grain consumed daily per head in experiment C was as follows: High, 0.2396 pound; low, 0.0816 pound; average, 0.1480 pound.

EXPERIMENT D.

The stock in experiment D was of poorer quality than that at any of the other stations, and the loss from sickness and deaths during the feeding period was so large as to affect the gains adversely in most of the lots. Feeding station No. 2 was used, the equipment of which has already been described. The methods of handling and feeding and the ration used were similar to those described under experiments A and C, except that 6 per cent of shorts replaced that much corn meal until about the middle of October, when the shorts were gradually dropped from the ration. Another difference was that condensed buttermilk, diluted with water, was used in mixing the feed. The supply of milk was short during a considerable part of the feeding season and water was used freely, one feed often being mixed entirely with water. This probably partially accounts for the poorer results obtained as compared with experiment A, where, in general, the feed and methods were similar. Another factor which helps to account for the difference in results was the condition of the feeding station where the experiment was conducted. As mentioned previously in describing the house in detail, this station was very drafty in cool or cold weather, and the birds were, in consequence, subject to colds. It is impossible to tell just how much influence the condition of the house had in producing sickness, but much of it was undoubtedly caused by drafts on the birds. The station was devoted entirely to feeding, the birds being disturbed only at feeding times and when the batteries were cleaned.

A large number of chickens in these lots had their wings broken before they reached the person who graded the dressed poultry. If the wings were broken before the bird was bled, the blood would clot at the broken spot and thus injure the quality of the flesh and cause the bird to be put in a lower grade. A brief survey of the existing conditions confirmed the statement previously noted that the quality of the stock when received was poorer than that received at the other stations. These broken wings were only in rare cases noted at the other stations. The care used in handling appeared to be as good as that in either experiment A or experiment C. This condition was not extensive enough, however, to allow a selection of lots for experimental work on this point. The addition of ground bone or some kind of meat food might prevent the bones from becoming so brittle, but the principal remedy seems to lie in a change of management. The addition of bone to the ration in a few batteries did not appear to affect the gains either way.

TABLE 7.—*Per cent gains of chickens in experiment D, by months and by length of feeding period in each month.*

Month.	Number of head.	Days fed.	Per cent gain.		
			High.	Low.	Average.
			Per cent.	Per cent.	Per cent.
August	1,261	16	38	28	32.3
Do	3,091	15	39	27	34.7
Do	788	14	32	32	32.0
Do	314	8	23	12	17.5
Do	309	7	22	9	13.3
Do	1,642	6	18	6	11.9
September	1,495	16	32	28	29.7
Do	7,087	15	39	23	30.4
Do	1,952	9	16	13	14.5
Do	10,717	8	23	2	12.6
Do	3,480	7	18	9	13.3
October	3,860	15	56	22	33.2
Do	386	14	31	22	26.5
Do	3,170	9	18	11	14.3
Do	24,526	8	16	4	10.5
November	2,285	15	34	10	21.7
Do	236	14	24	24	24.0
Do	22,720	8	35	5	10.0

TABLE 8.—*Summary of experiment D, arranged according to length of feeding period.*

Number of head.	Days fed.	Per cent gain.		
		High.	Low.	Average.
		Per cent.	Per cent.	Per cent.
2,756	16	38	28	31.0
16,323	15	56	10	30.0
1,410	14	32	22	27.5
5,122	9	18	11	14.4
58,277	8	35	2	12.7
3,789	7	22	9	13.3
1,642	6	18	6	11.9
89,319	....	56	2	20.1

The average daily grain consumption per head in this experiment was as follows: High, 0.2677 pound; low, 0.0966 pound; average, 0.1754 pound.

COMPARISON OF EXPERIMENTS C AND D.

The summaries of experiments C and D confirm in general the conclusions drawn from experiments A and B in respect to the gains, but they also tend to show that the 14 and 15 day feeding periods are profitable. A comparison of the gains made in these experiments with the gains in experiment A of lots fed the same length of time shows that the gains secured in experiment A were much better, although the ration was quite similar, but the relative conditions at the stations where the experiments were conducted would largely account for this difference. This would show that the conditions at the feeding station had as much influence as the feed in producing economical gains. The records of experiments C and D also show the wide variation in results secured in feeding. The average daily grain consumption for the season in experiment D is considerably greater than in experiment C, while the average gain in these experiments is nearly equal. This difference may be partly explained by the unfavorable housing conditions and the larger per cent of deaths in experiment D. As previously stated, the station in which the birds in experiment D were fed was cold and subject to drafts in the late fall.

AVERAGE DAILY CONSUMPTION OF GRAIN PER HEAD.

Table 9 shows the average daily consumption of grain (not including milk) per bird in experiments C and D. The average grain daily in experiments A and B is found in Tables I and II of the Appendix, but in these cases the average was for the period during which the lot was on feed, and therefore does not give the daily fluctuations shown in Table 9. The milk is added in getting the cost of feed per pound of gain. This table shows how much variation there is in the amount of feed which the birds will eat from day to day. This variation is affected by the weather, by the condition of the birds, and by the method of management, as well as by the kind of food. This average daily consumption is reported to headquarters each day and serves as a close check on the success which is being obtained by the feeder, as the average gains obtained at the station vary in accordance with the amount of feed consumed.

CONSUMPTION OF GRAIN.

TABLE 9.—*Average daily consumption of grain per head.*

EXPERIMENT C.

Date.	Amount consumed.	Date.	Amount consumed.	Date.	Amount consumed.	Date.	Amount consumed.
	Pounds.		*Pounds.*		*Pounds.*		*Pounds.*
August 16.....	0.1178	September 11.	0.1108	October 6.....	0.1418	October 31....	0.1774
August 17.....	.1130	September 12.	.1162	October 7.....	.1439	November 1..	.1836
August 18.....	.1380	September 13.	.1077	October 8.....	.1458	November 2..	.1949
August 19.....	.1311	September 14.	.1246	October 9.....	.1426	November 3..	.2016
August 20.....	.1326	September 15.	.1541	October 10....	.1647	November 4..	.1916
August 21.....	.1122	September 16.	.1075	October 11....	.1574	November 5..	.1883
August 22.....	.0988	September 17.	.1200	October 12....	.1532	November 6..	.1847
August 23.....	.0886	September 18.	.1080	October 13....	.1539	November 7..	.2158
August 24.....	.1138	September 19.	.1188	October 14....	.1479	November 8..	.1898
August 25.....	.0816	September 20.	.1440	October 15....	.1475	November 9..	.2120
August 26.....	.1129	September 21.	.1156	October 16....	.1612	November 10.	.1780
August 27.....	.1249	September 22.	.1067	October 17....	.1669	November 11.	.1898
August 28.....	.1250	September 23.	.0922	October 18....	.1650	November 12.	.1914
August 29.....	.1031	September 24.	.0913	October 19....	.1677	November 13.	.1718
August 30.....	.1131	September 25.	.0841	October 20....	.1646	November 14.	.2159
August 31.....	.1102	September 26.	.1033	October 21....	.1801	November 15.	.2082
September 1..	.1109	September 27.	.1266	October 22....	.1687	November 16.	.1734
September 2..	.0890	September 28.	.1381	October 23....	.1864	November 17.	.1738
September 3..	.1367	September 29.	.1329	October 24....	.1912	November 18.	.2396
September 4..	.0878	September 30.	.1255	October 25....	.1848	November 19.	.1870
September 5..	.1064	October 1.....	.1247	October 26....	.1912	November 20.	.2142
September 6..	.1025	October 2.....	.1297	October 27....	.1746	November 21.	.1689
September 7..	.1326	October 3.....	.1597	October 28....	.2035	November 22.	.1773
September 8..	.1074	October 4.....	.1576	October 29....	.1984	November 23.	.1930
September 9..	.1387	October 5.....	.1409	October 30....	.1737	November 24.	.1661
September 10.	.1125						

EXPERIMENT D.

Date.	Amount consumed.	Date.	Amount consumed.	Date.	Amount consumed.	Date.	Amount consumed.
	Pounds.		*Pounds.*		*Pounds.*		*Pounds.*
August 16.....	0.1517	September 11.	0.1220	October 6.....	0.2087	October 31....	0.2287
August 17.....	.1440	September 12.	.1755	October 7.....	.1839	November 1..	.2145
August 18.....	.1758	September 13.	.1810	October 8.....	.1918	November 2..	.1921
August 19.....	.1729	September 14.	.1543	October 9.....	.1704	November 3..	.1906
August 20.....	.2059	September 15.	.1726	October 10....	.2003	November 4..	.2233
August 21.....	.1517	September 16.	.1822	October 11....	.1561	November 5..	.1818
August 22.....	.1249	September 17.	.1804	October 12....	.1703	November 6..	.1741
August 23.....	.0982	September 18.	.1424	October 13....	.1612	November 7..	.2442
August 24.....	.0966	September 19.	.1623	October 14....	.1499	November 8..	.1830
August 25.....	.1306	September 20.	.1936	October 15....	.1674	November 9..	.1789
August 26.....	.1318	September 21.	.1328	October 16....	.1525	November 10.	.2483
August 27.....	.1373	September 22.	.1206	October 17....	.1814	November 11.	.1930
August 28.....	.1145	September 23.	.1624	October 18....	.1692	November 12.	.2012
August 29.....	.1368	September 24.	.1750	October 19....	.1553	November 13.	.2365
August 30.....	.1215	September 25.	.1103	October 20....	.1919	November 14.	.2677
August 31.....	.1604	September 26.	.1983	October 21....	.2079	November 15.	.2053
September 1..	.1183	September 27.	.1875	October 22....	.2600	November 16.	.2451
September 2..	.1383	September 28.	.1372	October 23....	.1724	November 17.	.2519
September 3..	.1207	September 29.	.1913	October 24....	.1587	November 18.	.2270
September 4..	.1185	September 30.	.1791	October 25....	.2362	November 19.	.1856
September 5..	.1862	October 1.....	.1499	October 26....	.1768	November 20.	.1962
September 6..	.1523	October 2.....	.1243	October 27....	.1522	November 21.	.2414
September 7..	.1229	October 3.....	.1624	October 28....	.2021	November 22.	.2288
September 8..	.1577	October 4.....	.1699	October 29....	.1908	November 23.	.2133
September 9..	.1532	October 5.....	.1893	October 30....	.1917	November 24.	.1965
September 10.	.1596						

DAILY DEATH RECORDS.

Table 10 gives the highest and average daily death records in percentages in each of the four experiments for the periods indicated. The remarks previously made concerning the stock under each experiment will help to explain the differences in this table. The total deaths as shown in the averages are small, but on very hot days the death record was high. October, as previously noted, is a bad month to feed on account of sickness and deaths among the birds. In experiment B the cripples were removed after the middle of September, which accounts for the low rate as compared to the other tables, where there were no cripples removed. These records confirm the statements made as to the health of the stock and the effect of the housing conditions at the various stations, allowing for the removal of the cripples in experiment B. A comparison of the death record with the percentage of gains in the respective feeding experiments shows that the death rate must be very low in order to get economical gains and that the gains vary inversely with the death rate.

TABLE 10.—*Daily death records.*

Date.	Experiment A.		Experiment B.		Experiment C.		Experiment D.	
	Highest.	Average.	Highest.	Average.	Highest.	Average.	Highest.	Average.
	Per cent.	*Per cent.*	*Per cent.*	*Per cent.*	*Per cent.*	*Per cent.*	*Per cent.*	*Per cent.*
July 27–Aug. 15	1.667	0.182	0.123	0.066				
Aug. 16–31	.161	.027	1.031	.166	0.262	0.108	0.222	0.132
September	.177	.038	.120	.059	.201	.039	.303	.125
October	.149	.023	.163	.061	.277	.087	.678	.210
Nov. 1–24	.463	.094	.207	.042	.373	.130	.449	.194

FATTENING HENS.

The accompanying table, No. 11, gives the results obtained in fattening a number of hens on trough feeding. Lots 1 to 6 were fed at station No. 3, and lots 7 to 14 at station No. 1. The main object of feeding was to increase the weight so as to get the hens into a higher grade, as very lightweight hens do not sell at a good price. In all these lots lightweight hens were selected and put on feed, while the heavier hens were killed without feeding any length of time. The hens to be killed without special feeding were given a full feed of corn chop and water and were held for several hours without feeding at one station before killing, but were killed with feed in their crops at another station, the food being removed after they were killed. If held without feeding and killed within a day after they reach the packing house, hens generally shrink from 1 to 3 per cent in weight. This shrinkage should be considered in calculating the profit or loss in feeding, as the hens which are fed for several days are weighed just before they are killed, while the hens which are not fed are credited with their weight as they reach the packing house.

Lots 1 to 6 were fed in various ways. The lots fed earliest in the season received cooked corn chop mixed with condensed buttermilk and water. After three or four days on this feed very bad cases of diarrhea developed, and the following lots were fed on a ration containing three-fourths corn chop and one-fourth low-grade wheat flour mixed with condensed buttermilk and water, which was scalded before being mixed with the grain. This ration kept the birds in better condition than the former feed, but did not entirely stop the diarrhea in the hens. Both the addition of low-grade flour and the scalding of the milk probably helped to stop the diarrhea. Ground bone was sprinkled lightly in the feed for a few of the batteries, and these batteries came through with little or no diarrhea when fed the ration containing the flour and scalded milk; but, as the table shows, the gains secured in all the lots were small and very variable, and the averages do not show any better gains from the second ration. Feeding the hens was then discontinued, partly because the gains cost too much and partly because there were not enough real light-weight hens to pay to feed. In selecting light hens a large number were apt to be Leghorns, which make poor gains.

Lots 7 to 14 received the same feed as the chickens in experiment B. The gains are very irregular, but a comparison with experiment B shows that it does not pay to cook the feed or make a separate mixture for the hens, as it is much easier to give the hens the same kind of feed which the chickens receive. The gains are greatly in favor of the uncooked regular feed, although most of the lots were fed longer in experiment B. A few hens off feed but not really sick may have been removed from the lots which were fed during October, so that the difference between the "Number in" and the "Number out" in these lots may not represent the dead in every case. The results at this station indicate that it pays to feed light hens, considering the increased value of the heavier birds, but apparently it is not always possible to get consistent gains. Hens do not eat their feed as quickly as springers and are harder to get on feed. If kept on feed for a long feeding period they must be watched carefully to prevent feather eating and other injurious habits. Better gains in hens are generally secured by cramming, but the labor is apt to make the total cost too high, and while some hens make very good gains when crammed, others fail to do anything.

TABLE 11.—*Results of fattening hens.*

STATION NO. 3.

Lot.	Number of birds.	Average weight in.	Dead.	Average weight out.	Dates fed.	Days fed.	Gain.	
							Pounds.	Per cent.
1	256	*Pounds.* 3.05	1	*Pounds.* 3.25	Oct. 25–Nov. 4....	11	47	6.00
2	256	2.93	3	3.13	Oct. 30–Nov. 8....	10	41	5.50
3	256	3.01	3	3.28	Oct. 13–19........	7	59	7.70
4	320	2.88	2	3.08	Oct. 8–14.........	7	58	6.30
5	320	2.91	4	3.03	Nov. 2–6..........	5	24	2.60
6	256	2.98	0	3.10	Oct. 17–21........	5	32	4.20

TABLE 11.—*Results of fattening hens*—Continued.

STATION NO. 1.

Lot.	Number of birds.	Average weight in.	Number out.[1]	Average weight out.	Dates fed.	Days fed.	Gain.	
							Pounds.	Per cent.
		Pounds.		*Pounds.*				
7	340	3.22	316	3.47	Sept. 29–Oct. 9...	11	32	2.00
8	340	3.46	336	4.06	Aug. 30–Sept. 9...	11	190	16.20
9	340	2.84	309	3.48	Oct. 2–12.........	11	207	21.00
10	340	3.59	314	3.68	Oct. 23–Nov. 1....	10	57	4.00
11	340	2.72	318	3.45	Oct. 30–Nov. 8....	10	241	26.00
12	340	3.37	309	3.74	Aug. 21–28........	8	9	.80
13	680	3.13	627	3.37	Oct. 13–19........	7	231	10.00
14	340	2.94	301	3.85	Oct. 17–23........	7	288	28.00

[1] Some hens were removed from these lots while on feed, so that the difference between the number in and the number out does not always represent dead birds.

SHRINKAGE IN DRESSING.

The birds were dressed and put into cold storage. The loss of weight or shrinkage in dressing (without drawing) for the different classes of birds varied as follows: Hens, 13.4 to 14.9 per cent, average 14.4 per cent; roasters, 13.7 to 16 per cent, average 14.7 per cent; springs, 9 to 14.5 per cent, average 12.1 per cent; and broilers, 14 to 14.7 per cent, average 14.3 per cent. The chickens at the different stations were fed a mixture of fine sand and very thin feed, or were first given a light feed and then sand and water for the last feed of the day before they were killed. Several of the pickers claimed that the chickens picked easier if the birds were watered freely before killing, and this practice was prevalent, although there was some difference of opinion as to the effect of the water.

CLEANING AND SPRAYING THE BATTERIES.

Where chickens are kept in large numbers cleanliness is a very important factor. Great care was exercised at all of the feeding stations to keep the batteries clean. In some stations the batteries were cleaned every day, in others, every other day, the latter plan prevailing at the majority of the stations. The droppings were scraped from the trays with a tin scraper. A scraper a trifle over half the width of the tray saves time in cleaning, if holding it does not tire the operator too much. Some prefer a narrower scraper. Many feeders dust air-slaked lime on each tray after cleaning. Another good method, which takes less time, is to spray the tray lightly with a hand sprayer after it is put back into the battery, using a coal-tar disinfectant. The batteries are generally sprayed with whitewash after each lot has been removed, although sometimes this was neglected and the batteries were only sprayed about once a month. The frequency of cleaning depends somewhat on the length of the feeding period, but batteries should be sprayed at least twice a month. Lime keeps the insect pests away.

POULTRY MANURE.

A large amount of poultry manure is produced daily at the feeding stations, and this has considerable fertilizing value, but most of the managers find it a source of extra expense instead of revenue. At one station the droppings were washed through a sewer into a river, which involved considerable labor; at other places the manure was loaded into an empty wagon kept for that purpose, which was hauled away each day. In some localities in the Middle West the packers can get farmers to remove the manure without paying for the labor of hauling, but as the feeding stations are usually located in the towns or cities the manure must be removed regularly and promptly, or it becomes a public nuisance. In such cases the manure should be taken away daily, and farmers are apt to be irregular about doing the work. Various methods of drying the droppings and putting them on the market for fertilizer have been suggested and tried on a small scale, but apparently without success. The farmers in this section will learn the value and necessity of using manure on their land in time, and will then be glad to avail themselves of a daily supply of poultry droppings for nothing.

Some of the packers hope to operate farms in connection with their feeding stations, thus utilizing the manure to good advantage. Rightly handled, a farm should be a profitable undertaking in connection with a feeding station, not only utilizing the various by-products, but also as a breeding establishment for poultry, where the packer could raise purebred stock of the best market types to distribute in the territory from which he draws his products. This distribution could be done either at a nominal price or in exchange for the same weight of undesirable types of male birds of mixed breeding. The packer could also select the best pullets from the stock shipped in to be marketed and keep them for spring egg production, killing these pullets at the end of their laying period, when they would be worth about as much for dressed poultry as they had cost in the fall or winter.

Records of the amount of droppings from the fattening stock kept at various times during the feeding season showed that the number of pounds produced per 100 birds varied from day to day. One hundred "springs" averaged 11 pounds of manure daily, which would mean over half a ton of manure a day for each 10,000 chickens on feed. The consistency of the droppings is a fair indication of the condition of the bowels of the chickens. The droppings should be soft but not watery. Confinement and sour milk make softer droppings than are obtained from poultry kept on the range and fed on whole or ground grains. The feeder should observe the droppings occasionally and feed accordingly.

In case the birds have excessive diarrhea, it is a good plan to scald the milk for one or two feedings, which will generally bring the birds back to normal condition.

KEEPING RECORDS.

Careful records were kept of all the operations in the packing houses, for which purpose various systems of checking and rechecking the different lots of chickens were used. The person in charge of each room or branch of the work made reports to the office covering the work done in his room for each day. The packers, of course, aim to systematize labor, save any wastes for by-products, and reduce cost in every line, at the same time improving the quality of the product. A good type of feeding-station report is shown below. Careful records were kept of the cost of producing gains and of killing and dressing birds, while some of the packers had elaborate records which showed all the expenses incurred by a lot of chickens until they went into storage. By a careful study of these records the manager of the packing house could tell what his product was costing, and could figure out how the cost was divided. There are many different ideas as to the best way to keep such records, but the object should be to show the cost of each part of the work accurately, yet as simply as possible.

FEEDING-STATION REPORT.

Date_____, 191 .

Total number of chickens on feed to-day_____

FEED.

Fed to-day.	Weight.	Price.	Total cost.	Weight fed per 100 head.
Corn meal..pounds..				
Oatmeal...do....				
Milk...gallons..				
Tallow..pounds..				
Meat...do....				
Grit..				
Total...				

COST OF FEED PER 100 HEAD.

Cost of feed per 100 head...	$............
Cost of labor per 100 head..	....
Total average cost per 100 head..	....
Total cost labor to-day...	....

CHICKENS PUT ON FEED.

	Head.	Weight.	Average weight.
	Number.	*Pounds.*	*Pounds.*
Put on feed to-day.............................			
Put on feed to-day.............................			
Put on feed to-day.............................			
Put on feed to-day.............................			

CHICKENS KILLED.

	Days fed.	Head.	Weight.	Gain.	Per cent gain.	Died.
		Number.	*Pounds.*	*Pounds.*	*Per cent.*	*Number.*
Killed to-day						
Killed to-day						
Killed to-day						
Killed to-day						

Sick to-day...............head. Weight..............pounds. Average weight.........pounds.
Dead to-day...............head. Weight..............pounds. Average weight.........pounds.

(Signed) _____, *Manager.*

CONCLUSIONS.

1. The Plymouth Rock and other varieties of general-purpose fowls make more economical gains in fattening than the Mediterranean class, such as Leghorns.

2. Chickens of the same breed vary greatly in their ability to put on flesh. This variation may lead to gross error in drawing conclusions from experiments in feeding poultry which deal with only a small number of birds.

3. Muslin or duck cloth can be used to good advantage to replace the windows or part of the walls of feeding stations.

4. If a feeding station is properly constructed, good ventilation can be secured without having a large open space in the top of the building, such as a monitor top. Such buildings can be constructed more cheaply than those with a large amount of air space per bird, by using muslin curtains for the walls.

5. The use of portable feeding batteries is more easily adapted to varying conditions, involves less labor, and turns the birds out in better condition than the stationary batteries.

6. Low-grade wheat flour is a more economical feed than oat flour in fattening rations for chickens at the present prices of grain.

7. The average person will get better results in fattening by feeding three times rather than twice daily.

8. The amount of grain required to produce a pound of flesh in fattening chickens varied in experiment A from 1.92 to 5.35 pounds, with an average of 3.26 pounds; while in experiment B the amount varied from 1.29 to 8.45 pounds, with an average of 3.26 pounds.

The total cost of feed per pound of gain varied from 3.71 to 10.37 cents, and averaged 6.45 cents in experiment A, while in experiment B the cost varied from 3.15 to 19.90 cents, and averaged 7.74 cents.

The cost of labor for a pound of gain in flesh varied from 0.88 to 2.81 cents and averaged 1.40 cents in experiment A, while in experiment B the cost varied from 1.14 to 5.63 cents, and averaged 2.59 cents.

The cost of both feed and labor to produce a pound of gain in fattening varied from 4.61 to 13.14 cents, and averaged 7.85 cents in experiment A; and it varied from 4.35 to 27.20 cents, and averaged 10.33 cents in experiment B.

The average total cost of feed and labor per pound of gain for all the birds in experiments A and B was 9.09 cents; the average cost of feed alone, 7.10 cents.

9. The cheaper gains were made in the shorter feeding periods (7 or 8 days) and by the light chickens.

10. Hens make poorer gains than chickens in crate fattening. Fattening hens by this method is profitable only under certain conditions.

APPENDIX.

TABLE I.—Details of feeding experiment A.

Lot.	Class.	Number in.	Total weight.	Average weight.	Dates fed.	Days fed.	Total feed.	Number out.[1]	Total weight.	Average weight.	Total gain.	Per cent gain.	Average number fed.[2]	Average grain daily per head.	Grain per pound of gain.	Total cost of labor.	Cost of labor per pound of gain.	Cost of feed per pound of gain.[3]	Total cost per pound of gain.[3]
		Head.	Pounds.	Pounds.			Pounds.	Head.	Pounds.	Pounds.	Pounds.	Per ct.	Head.	Pounds.	Pounds.	Dollars.	Cents.	Cents.	Cents.
1	Springers	480	900	1.88	Aug. 6 to 15	10	706.0	477	1,093	2.29	193	21.4	478.5	0.1476	3.66	3.49	1.81	7.06	8.87
2	...do...	240	481	2.00	Aug. 7 to 15	9	317.0	238	556	2.34	75	15.6	239.0	.1476	4.23	1.57	2.09	8.16	10.25
3	...do...	160	312	1.95	Aug. 10 to 15	6	138.0	159	341	2.14	29	9.3	159.5	.1442	4.76	.70	2.41	9.19	11.60
4	...do...	400	771	1.93	Aug. 10 to 16	7	400.0	395	859	2.17	88	11.4	397.5	.1436	4.55	2.03	2.31	8.78	11.09
5	...do...	640	1,212	1.89	Aug. 12 to 17	6	530.0	633	1,311	2.07	99	8.2	636.5	.1389	5.35	2.78	2.81	10.33	13.14
6	...do...	240	435	1.81	Aug. 14 to 22	9	321.0	239	541	2.26	106	24.4	239.5	.1491	3.03	1.57	1.48	5.85	7.33
7	...do...	320	620	1.94	Aug. 16 to 23	8	377.0	318	733	2.31	113	18.2	319.0	.1478	3.34	1.86	1.65	6.45	8.10
8	...do...	480	972	2.03	Aug. 17 to 24	8	566.0	479	1,147	2.39	175	18.0	479.5	.1475	3.23	2.80	1.60	6.23	7.83
9	...do...	160	308	1.93	Aug. 18 to 25	8	191.0	159	358	2.25	50	16.2	159.5	.1498	3.82	.93	1.86	7.37	9.23
10	...do...	240	520	2.17	Aug. 19 to 28	10	367.0	239	642	2.69	122	23.5	239.5	.1532	3.01	1.75	1.43	5.81	7.24
11	...do...	640	1,224	1.91	Aug. 20 to 28	9	876.0	638	1,469	2.30	245	20.0	639.0	.1524	3.58	4.19	1.71	6.91	8.62
12	...do...	480	910	1.90	Aug. 21 to 29	9	630.0	480	1,140	2.38	230	25.3	480.0	.1458	2.74	3.15	1.37	5.29	6.66
13	...do...	400	756	1.89	Aug. 23 to 31	9	521.0	398	928	2.33	172	22.8	399.0	.1452	3.03	2.62	1.52	5.85	7.37
14	...do...	320	714	2.23	Aug. 24 to Sept. 1	9	440.0	319	875	2.74	161	22.5	319.5	.1530	2.73	2.10	1.30	5.27	6.57
15	...do...	160	307	1.92	Aug. 26 to Sept. 1	7	175.0	160	398	2.49	91	29.6	160.0	.1560	1.92	.82	.90	3.71	4.61
16	...do...	800	1,646	2.06	Aug. 27 to Sept. 2	7	883.0	798	1,939	2.43	293	17.8	799.0	.1578	3.01	4.08	1.39	5.81	7.20
17	...do...	400	785	1.96	Aug. 28 to Sept. 5	9	558.0	397	968	2.44	183	23.3	398.5	.1556	3.05	2.61	1.43	5.89	7.32
18	...do...	240	526	2.19	Aug. 30 to Sept. 6	8	296.0	237	618	2.61	92	17.5	238.5	.1551	3.22	1.39	1.51	6.21	7.72
19	...do...	720	1,575	2.19	Aug. 31 to Sept. 6	7	794.0	717	1,916	2.67	341	21.7	718.5	.1578	2.33	3.67	1.08	4.50	5.58
20	...do...	1,440	3,205	2.23	Sept. 1 to 7	7	1,606.0	1,436	3,746	2.61	541	16.9	1,438.0	.1596	2.97	7.34	1.36	5.73	7.09
21	...do...	800	1,865	2.33	Sept. 1 to 9	9	1,110.0	794	2,161	2.72	296	15.9	797.0	.1548	3.75	5.23	1.77	7.73	9.50
22	...do...	480	1,006	2.10	Sept. 2 to 9	8	580.0	478	1,230	2.57	224	22.3	479.0	.1513	2.59	2.79	1.25	5.34	6.59

[1] The difference between the "Number in" and the "Number out" represents the dead birds in this table. Such is not the case with Table II, where the "cripples" were removed during the feeding period and their dressed weight credited to the lot when it was killed.
[2] For the purpose of computing results, the "Average number fed" is arrived at by adding the "Number in" and the "Number out" and dividing by 2, thus getting an average.
[3] The cost of feed per pound of gain includes the cost of the milk, which is added to the cost of the grain.

56 FATTENING POULTRY.

TABLE I.—*Details of feeding experiment A*—Continued.

Lot.	Class.	Number in.	Total weight.	Average weight.	Dates fed.	Days fed.	Total feed.	Number out.	Total weight.	Average weight.	Total gain.	Per cent gain.	Average number fed.	Average grain daily per head.	Grain per pound of gain.	Total cost of labor.	Cost of labor per pound of gain.	Cost of feed per pound of gain.	Total cost per pound of gain.
		Head.	Pounds.	Pounds.			Pounds.	Head.	Pounds.	Pounds.	Pounds.	Per ct.	Head.	Pounds.	Pounds.	Dollars.	Cents.	Cents.	Cents.
23	Springers	1,200	2,686	2.24	Sept. 7 to 13	7	1,391.0	1,200	3,156	2.63	470	17.5	1,200.0	0.1656	2.96	6.12	1.30	6.10	7.40
24	...do...	480	1,049	2.19	Sept. 8 to 15	8	648.0	478	1,236	2.59	187	17.8	479.0	.1662	3.47	2.79	1.49	7.15	8.64
25	...do...	720	1,681	2.33	Sept. 9 to 15	7	877.0	718	2,012	2.80	331	19.7	719.0	.1743	2.65	3.67	1.11	5.46	6.57
26	...do...	720	1,693	2.35	Sept. 10 to 16	7	910.0	719	2,024	2.80	331	19.6	719.5	.1806	2.75	3.67	1.11	5.67	6.78
27	...do...	640	1,525	2.38	Sept. 11 to 18	8	875.0	635	1,776	2.80	251	16.5	637.5	.1715	3.49	3.72	1.48	7.19	8.67
28	...do...	480	1,101	2.29	Sept. 13 to 18	6	500.0	479	1,220	2.55	119	10.8	479.5	.1739	4.20	2.10	1.76	8.65	10.41
29	...do...	1,040	2,396	2.30	Sept. 14 to 19	6	1,060.0	1,029	2,779	2.70	383	16.0	1,034.5	.1708	2.77	4.52	1.18	5.71	6.89
30	...do...	960	2,163	2.25	Sept. 16 to 21	6	940.0	958	2,506	2.62	343	15.9	959.0	.1634	2.74	4.19	1.22	5.64	6.86
31	...do...	1,200	2,848	2.37	Sept. 17 to 23	6	1,291.0	1,194	3,375	2.83	527	18.5	1,197.0	.1541	2.45	6.11	1.16	5.05	6.21
32	...do...	880	2,049	2.33	Sept. 18 to 25	8	1,071.0	875	2,387	2.73	338	16.5	877.5	.1526	3.17	5.12	1.51	6.53	8.04
33	...do...	400	960	2.40	Sept. 20 to 26	7	442.0	400	1,139	2.85	179	18.6	400.0	.1578	2.47	2.04	1.14	5.09	6.23
34	...do...	400	886	2.22	Sept. 19 to 25	7	431.0	398	1,096	2.75	210	23.7	399.0	.1644	2.05	2.04	.97	4.22	5.19
35	...do...	1,600	3,778	2.36	Sept. 22 to 28	7	1,817.0	1,592	4,623	2.90	845	22.4	1,596.0	.1626	2.15	8.14	.96	4.43	5.39
36	...do...	480	1,204	2.51	Sept. 23 to 29	7	550.0	477	1,414	2.96	210	17.4	478.5	.1638	2.62	2.44	1.16	5.40	6.56
37	...do...	640	1,488	2.33	Sept. 24 to Oct. 2	9	958.0	639	1,856	2.90	368	24.7	639.5	.1664	2.60	4.20	1.14	5.36	6.50
38	...do...	1,040	2,729	2.62	Sept. 25 to Oct. 3	9	1,533.0	1,032	3,172	3.07	443	16.2	1,036.0	.1644	3.46	6.80	1.53	6.75	8.28
39	...do...	640	1,533	2.40	Sept. 28 to Oct. 3	6	610.0	636	1,818	2.86	285	18.6	638.0	.1594	2.14	2.79	.98	4.17	5.15
40	...do...	800	2,067	2.58	Sept. 29 to Oct. 5	7	927.0	792	2,530	3.19	463	23.6	796.0	.1662	2.00	4.06	.88	3.90	4.78
41	...do...	1,360	3,198	2.35	Sept. 29 to Oct. 6	8	1,880.0	1,358	4,064	2.99	866	27.1	1,359.0	.1726	2.17	7.93	.92	4.23	5.15
42	...do...	800	2,018	2.52	Oct. 2 to 10	9	1,342.0	792	2,545	3.21	527	26.1	796.0	.1873	2.55	5.22	.99	4.97	5.96
43	...do...	800	2,068	2.59	Oct. 5 to 11	7	1,107.0	798	2,388	2.99	320	15.5	799.0	.1960	2.46	4.08	1.28	6.75	8.03
44	...do...	800	2,104	2.63	Oct. 6 to 12	7	1,073.0	797	2,522	3.16	418	19.9	798.5	.1921	2.57	4.07	.97	5.01	5.98
45	...do...	480	1,265	2.64	Oct. 7 to 13	7	645.0	480	1,497	3.12	232	18.3	480.0	.1918	2.78	2.45	1.06	5.42	6.48
46	...do...	720	1,958	2.72	Oct. 9 to 17	9	1,249.0	716	2,370	3.31	412	21.0	718.0	.1932	3.03	4.71	1.14	5.91	7.05
47	...do...	560	1,543	2.76	Oct. 11 to 18	8	857.0	560	1,783	3.20	250	16.2	560.0	.1912	3.43	3.27	1.31	6.69	8.00
48	...do...	640	1,680	2.62	Oct. 12 to 18	7	873.0	639	2,001	3.13	321	19.1	639.5	.1949	2.72	3.26	1.02	5.30	6.32
49	...do...	1,200	3,406	2.84	Oct. 15 to 21	7	1,671.0	1,195	4,049	3.39	643	18.9	1,197.5	.1993	2.60	6.11	.95	5.07	6.02
50	...do...	800	2,197	2.75	Oct. 15 to 21	7	1,152.0	797	2,514	3.15	317	14.4	798.5	.2060	3.63	4.07	1.28	7.08	8.36
51	...do...	880	2,429	2.76	Oct. 16 to 23	8	1,454.0	874	2,938	3.36	509	20.5	877.0	.2073	2.86	5.11	1.00	6.58	6.58
52	...do...	960	2,991	3.12	Oct. 19 to 26	8	1,622.0	955	3,440	3.60	449	15.0	957.5	.2117	3.61	5.58	1.24	7.04	8.28
53	...do...	640	1,945	3.04	Oct. 25 to Nov. 1	8	1,086.0	639	2,298	3.60	353	18.1	639.5	.2024	2.93	3.73	1.06	5.71	6.77

54	...do.....	800	2,179	2.72	Oct. 26 to Nov. 2.	8	1,270.0	797	2,504	3.14	325	14.9	798.5	.2001	3.94	4.66	1.43	7.68	9.11
55	...do.....	720	2,322	3.23	Oct. 30 to Nov. 8.	10	1,425.0	713	2,590	3.63	268	11.5	716.5	.1988	5.32	5.22	1.95	10.37	12.32
56	...do.....	640	1,892	2.96	Oct. 30 to Nov. 7.	9	1,149.0	634	2,216	3.50	324	17.1	637.0	.2004	3.55	4.18	1.29	6.92	8.21
57	...do.....	640	1,997	3.12	Nov. 1 to 8....	8	1,009.0	632	2,293	3.63	296	14.8	636.0	.1982	3.41	3.71	1.25	6.65	7.90
58	...do.....	640	1,774	2.77	Nov. 4 to 13...	10	1,293.0	627	2,085	3.33	311	17.6	633.5	.2040	4.16	4.62	1.49	8.11	9.60
59	...do.....	1,200	3,807	3.17	Nov. 6 to 14...	9	2,161.0	1,188	4,246	3.57	439	11.5	1,194.0	.2001	4.92	7.83	1.78	9.59	11.37
60	...do.....	880	2,611	2.97	Nov. 3 to 11...	9	1,601.0	863	2,925	3.39	314	12.0	871.5	.2041	5.10	5.72	1.82	9.95	11.77
61	...do.....	960	2,960	3.08	Nov. 10 to 17..	8	1,584.0	941	3,320	3.53	360	12.2	950.5	.2062	4.40	5.54	1.54	8.58	10.12
62	...do.....	1,440	4,873	3.38	Nov. 11 to 18..	8	2,275.0	1,422	5,406	3.80	533	10.9	1,431.0	.1987	4.27	8.35	1.57	8.33	9.90
63	...do.....	960	3,001	3.13	Nov. 13 to 21..	9	1,683.0	953	3,338	3.50	337	11.2	956.5	.1955	4.99	6.28	1.86	9.73	11.59

58 FATTENING POULTRY.

TABLE II.—*Details of feeding experiment B.*

Lot.	Class.	Number in.	Total weight.	Average weight.	Dates fed.	Days fed.	Total feed.	Number out.[1]	Total weight.	Average weight.	Total gain.	Per cent gain.	Average number fed.[3]	Dead.	Average grain daily per head.	Grain per pound of gain.	Total cost of labor.	Cost of labor per pound of gain.	Cost of feed per pound of gain.	Total cost per pound of gain.
		Head.	*Pounds.*	*Pounds.*			*Pounds.*	*Head.*	*Pounds.*	*Pounds.*	*Lbs.*	*Per ct.*	*Head.*		*Pounds.*	*Pounds.*	*Dollars.*	*Cents.*	*Cents.*	*Cents.*
1	Springers	450	835	1.86	July 28 to Aug. 10.	14	865.3	442	1,154	2.61	319	38.2	446.0	5	0.1386	2.71	5.80	1.84	6.26	8.10
2	...do......	450	858	1.91	July 29 to Aug. 11.	14	819.5	435	1,164	2.68	306	35.7	442.5	3	.1338	2.68	5.85	1.91	6.19	8.10
3	...do......	450	744	1.65	July 30 to Aug. 12.	14	799.8	449	1,174	2.61	430	57.8	449.5	1	.1271	1.86	5.94	1.38	4.30	5.68
4	...do......	450	817	1.82	Aug. 3 to 16	14	804.7	444	1,099	2.48	282	34.5	447.0	6	.1286	2.85	5.91	2.10	6.58	8.68
5	...do......	1,130	2,300	2.04	Aug. 4 to 17	14	2,045.0	1,119	3,084	2.76	784	34.1	1,124.5	11	.1299	2.61	14.86	1.90	6.03	7.93
6	...do......	1,125	2,186	1.94	Aug. 5 to 18	14	2,024.0	1,120	3,008	2.69	822	37.6	1,122.5	11	.1288	2.46	14.83	1.80	5.68	7.48
7	...do......	1,350	2,588	1.92	Aug. 6 to 19	14	2,455.0	1,348	3,617	2.68	1,029	39.8	1,349.0	8	.1300	2.39	17.83	1.73	5.52	7.25
8	...do......	900	1,712	1.90	Aug. 7 to 21	15	1,713.0	885	2,375	2.68	663	38.7	892.5	14	.1280	2.58	12.64	1.91	5.96	7.87
9	...do......	1,125	2,029	1.80	Aug. 8 to 15	8	1,159.0	1,107	2,668	2.41	637	31.4	1,116.0	15	.1298	1.82	8.43	1.33	4.20	5.52
10	...do......	1,125	2,325	2.07	Aug. 11 to 23	13	1,769.0	1,110	3,046	2.74	721	31.0	1,117.5	15	.1218	2.45	13.71	1.90	5.66	7.56
11	...do......	1,125	2,315	2.06	Aug. 12 to 24	13	1,727.0	1,110	2,933	2.64	618	26.7	1,117.5	14	.1189	2.79	13.71	2.22	6.44	8.66
12	...do......	1,125	2,319	2.06	Aug. 13 to 25	13	1,707.0	1,110	2,988	2.69	669	29.8	1,117.5	21	.1175	2.55	13.71	2.05	5.89	7.94
13	...do......	1,125	2,398	2.13	Aug. 14 to 26	13	1,687.0	1,054	2,800	2.66	402	16.8	1,089.5	6	.1191	4.20	13.37	3.33	9.70	13.03
14	...do......	1,125	2,499	2.22	Aug. 16 to 22	7	936.0	1,094	2,696	2.46	197	7.9	1,109.5	40	.1205	4.75	7.33	3.72	10.97	14.69
15	...do......	1,125	2,466	2.19	Aug. 19 to 28	10	1,323.0	1,110	2,992	2.70	526	21.3	1,117.5	6	.1184	2.52	10.55	2.01	5.82	7.83
16	...do......	1,125	2,554	2.27	Aug. 21 to 29	9	1,211.0	1,112	3,137	2.82	583	22.8	1,118.5	16	.1203	2.08	9.51	1.63	4.80	6.43
17	...do......	1,125	2,517	2.24	Aug. 22 to 30	9	1,246.0	1,114	3,121	2.80	604	24.0	1,119.5	13	.1237	2.06	9.50	1.57	4.76	6.33
18	...do......	1,125	2,447	2.18	Aug. 24 to 31	8	1,192.0	1,116	3,065	2.75	618	25.3	1,120.5	11	.1333	1.93	8.46	1.37	4.46	5.83
19	...do......	1,125	2,629	2.34	Sept. 2 to 9	8	1,322.0	1,130	3,305	2.92	541	20.6	1,127.5	7	.1466	2.44	8.51	1.57	5.64	7.21
20	...do......	1,125	3,607	2.53	Sept. 9 to 15	7	1,332.0	1,354	4,251	3.14	644	17.9	1,352.0	2	.1406	2.07	9.02	1.40	4.76	6.16
21	...do......	1,350	3,416	2.53	Sept. 10 to 16	7	1,391.0	1,353	4,210	3.11	793	23.1	1,351.5	12	.1467	1.75	9.02	1.15	4.03	5.17
22	...do......	1,350	3,509	2.60	Sept. 11 to 18	8	1,599.0	1,341	4,194	3.13	685	19.5	1,345.5	9	.1485	2.33	10.26	1.48	5.36	6.84
23	...do......	1,350	3,400	2.52	Sept. 12 to 21	10	1,924.0	1,340	4,307	3.21	907	26.7	1,345.0	10	.1430	2.12	12.82	1.41	4.88	6.29
24	...do......	1,350	3,666	2.72	Sept. 13 to 22	10	1,916.0	1,338	4,183	3.13	517	14.1	1,344.0	12	.1425	3.71	12.81	2.48	8.53	11.01
25	Broilers	450	820	1.82	Sept. 13 to 21	9	580.0	446	1,217	2.73	297	36.2	448.0	4	.1439	1.95	3.84	1.28	4.49	5.77
26	Roasters	1,350	4,031	2.99	Sept. 16 to 28	13	2,600.0	1,336	4,827	3.62	796	19.7	1,343.0	14	.1489	3.27	16.64	2.09	7.52	9.61
27	Broilers	900	2,204	2.45	Sept. 16 to 23	8	1,006.0	890	2,325	2.61	119	5.4	895.0	9	.1405	8.45	6.82	5.73	19.44	25.17
28	Roasters	900	2,806	3.12	Sept. 17 to 29	13	1,718.0	885	3,243	3.66	437	15.6	892.5	10	.1490	3.93	11.06	2.53	9.04	11.57
29	Broilers	450	1,009	2.38	Sept. 17 to 23	7	463.0	452	1,174	2.60	106	9.9	451.0	1	.1370	4.08	3.01	2.84	9.38	12.22
30	Roasters	900	2,794	3.10	Sept. 19 to Oct. 2	14	1,825.0	889	3,291	3.70	497	17.8	894.5	11	.1457	3.67	11.93	2.40	8.44	10.84
31	Broilers	450	1,065	2.37	Sept. 20 to 26	7	453.0	449	1,228	2.73	163	15.3	449.5	1	.1438	2.78	3.00	1.84	6.39	8.23
32	Roasters	900	2,755	3.06	Sept. 21 to Oct. 3	13	1,726.0	894	3,207	3.59	452	16.4	897.0	6	.1490	3.82	11.11	2.46	8.79	11.25

APPENDIX.

33	Broilers	900	2,003	2.23	Sept. 22 to 28	7	975.0	893	2,210	2.47	207	10.3	896.5	5	.1553	4.71	5.98	2.89	10.83	13.72
34	Roasters	450	1,510	3.36	Sept. 22 to Oct. 5.	14	925.0	440	1,721	3.91	231	15.3	445.0	3	.1485	4.00	5.94	2.57	9.20	11.77
35	...do...	900	2,954	3.28	Sept. 24 to Oct. 5.	12	1,615.0	891	3,285	3.69	341	11.5	895.5	5	.1503	4.74	10.24	3.00	10.90	13.90
36	Broilers	450	980	2.13	Sept. 25 to Oct. 2.	8	555.0	449	1,196	2.66	233	24.3	449.5	1	.1542	2.38	3.43	1.47	5.47	6.94
37	Roasters	900	2,555	2.84	Sept. 26 to Oct. 8.	13	1,668.0	880	3,278	3.73	749	29.3	890.0	15	.1542	2.27	11.03	1.47	5.22	6.69
38	Broilers	450	963	2.14	Sept. 27 to Oct. 2.	6	409.0	446	1,120	2.51	157	16.3	448.0	8	.1522	2.61	2.56	1.63	6.00	7.63
39	Roasters	800	2,525	3.16	Sept. 30 to Oct. 11.	12	1,270.0	724	2,674	3.69	337	13.3	762.0	7	.1389	3.77	8.71	2.58	8.67	11.25
40	...do...	800	2,742	3.43	Oct. 2 to 10	9	1,005.0	747	2,858	3.83	235	8.6	773.5	0	.1356	4.28	6.63	2.82	9.84	12.66
41	Broilers	900	2,084	2.32	Oct. 2 to 8	7	885.0	899	2,277	2.62	235	11.3	884.5	0	.1430	3.77	5.90	2.51	8.67	11.18
42	Roasters	800	2,747	3.43	Oct. 5 to 13	9	966.0	784	2,876	3.67	314	11.4	792.0	3	.1355	3.08	8.60	2.74	7.48	10.22
43	...do...	400	1,390	3.48	Oct. 6 to 14	9	457.0	360	1,312	3.64	105	7.6	380.0	1	.1337	4.35	4.12	3.92	10.57	14.49
44	Springers	680	2,113	3.11	Oct. 7 to 16	10	887.0	636	2,339	3.68	248	11.7	658.0	0	.1348	3.58	7.94	2.74	8.70	11.90
45	Broilers	450	994	2.21	Oct. 8 to 14	7	410.0	429	1,105	2.58	117	15.8	439.5	3	.1333	2.61	3.71	2.36	6.34	8.70
46	Roasters	800	2,919	3.65	Oct. 9 to 17	9	958.0	774	2,965	3.83	90	4.0	787.0	2	.1352	8.19	8.54	7.30	19.90	27.20
47	Broilers	450	1,168	2.60	Oct. 9 to 15	7	419.0	431	1,217	2.82	117	7.7	440.5	3	.1358	4.66	3.72	4.13	11.82	15.45
48	Springers	400	1,073	2.68	Oct. 12 to 18	7	366.0	387	1,481	3.83	144	13.4	393.5	2	.1329	2.54	3.32	2.31	6.17	8.48
49	...do...	336	1,283	3.82	Oct. 13 to 23	11	497.0	304	1,341	4.41	136	10.6	320.0	1	.1412	3.65	4.25	2.70	8.87	12.00
50	...do...	400	1,275	3.19	Oct. 14 to 23	10	542.0	357	1,310	3.67	169	13.3	378.5	6	.1432	3.21	4.56	3.13	7.80	10.50
51	...do...	400	1,334	3.34	Oct. 15 to 24	10	578.0	378	1,384	3.66	147	11.0	389.0	5	.1455	3.93	4.69	2.70	9.55	12.74
52	...do...	800	2,482	3.10	Oct. 18 to 25	8	945.0	723	2,620	3.62	317	12.8	761.5	3	.1550	2.98	7.35	2.32	7.24	9.56
53	...do...	800	2,652	3.32	Oct. 19 to 26	8	901.0	769	2,848	3.70	272	10.3	784.5	0	.1579	3.64	7.57	2.78	8.85	11.63
54	...do...	800	2,230	2.79	Oct. 21 to 27	7	901.0	749	2,424	3.24	358	16.1	774.5	6	.1662	2.52	6.42	1.79	6.12	7.91
55	Roasters	340	1,050	3.09	Oct. 21 to 30	10	543.0	312	1,399	4.48	141	13.4	326.0	2	.1666	3.85	3.93	2.79	9.36	12.15
56	Springers	400	1,129	2.82	Oct. 24 to 30	7	466.0	370	1,205	3.26	187	16.6	385.0	0	.1728	2.49	3.25	1.74	6.05	7.79
57	Roasters	336	1,391	4.14	Oct. 24 to Nov. 3.	11	604.0	309	1,345	4.35	93	6.7	322.5	0	.1702	6.49	4.28	4.60	15.77	20.37
58	Springers	800	2,172	2.72	Oct. 25 to 31	7	939.0	769	2,447	3.18	355	16.3	784.5	0	.1710	2.65	6.62	1.86	6.44	8.30
59	Roasters	400	1,365	3.41	Oct. 26 to Nov. 3.	10	613.0	322	1,433	4.45	105	7.7	361.0	0	.1698	5.84	4.35	4.14	14.19	18.33
60	...do...	672	2,612	3.89	Oct. 28 to Nov. 6.	10	1,100.0	614	2,730	4.45	317	12.1	643.0	2	.1711	3.47	7.75	2.44	8.43	10.87
61	Springers	800	2,125	2.66	Oct. 28 to Nov. 4.	8	1,020.0	718	2,389	3.33	385	18.1	759.0	2	.1680	2.65	7.32	1.90	6.44	8.34
62	Roasters	672	2,703	4.02	Oct. 30 to Nov. 8.	10	1,034.0	589	2,722	4.62	285	10.5	630.5	2	.1639	3.63	7.60	2.67	8.82	11.49
63	Springers	800	1,930	2.41	Oct. 30 to Nov. 5.	7	965.0	785	2,446	3.12	558	28.4	792.5	1	.1739	1.73	6.69	1.20	4.20	5.40
64	Broilers	400	1,090	2.73	Nov. 2 to 9	8	474.0	384	1,282	3.34	229	21.0	392.0	0	.1510	2.07	4.90	2.14	5.05	7.19
65	Roasters	672	2,598	3.87	Nov. 2 to 10	9	903.0	663	2,661	4.01	188	7.2	667.5	2	.1502	4.80	9.39	4.99	11.71	16.70
66	Broilers	400	965	2.41	Nov. 4 to 10	7	415.0	386	1,151	2.98	258	26.7	393.0	1	.1508	1.61	4.30	1.67	3.93	5.60
67	Roasters	672	2,507	3.73	Nov. 5 to 13	9	892.0	606	2,673	4.41	274	10.9	639.0	7	.1550	3.26	8.99	3.28	7.95	11.23

[1] The difference between the "Number in" and the "Number out" does not represent the dead birds in this table, as the "cripples" were removed during the feeding period and their dressed weight credited to the lot when it was killed.

[2] For the purpose of computing results, the "Average number fed" is arrived at by adding the "Number in" and the "Number out" and dividing by 2, thus getting an average.

60 FATTENING POULTRY.

TABLE II.—*Details of feeding experiment B—Continued.*

Lot.	Class.	Number in.	Total weight.	Average weight.	Dates fed.	Days fed.	Total feed.	Number out.	Total weight.	Average weight.	Total gain.	Per cent gain.	Average number fed.	Dead.	Average grain daily per head.	Grain per pound of gain.	Total cost of labor.	Cost of labor per pound of gain.	Cost of feed per pound of gain.	Total cost per pound of gain.
		Head.	Pounds.	Pounds.			Pounds.	Head.	Pounds.	Pounds.	Lbs.	Per ct.	Head.		Pounds.	Pounds.	Dollars.	Cents.	Cents.	Cents.
68	Broilers	400	941	2.35	Nov. 6 to 13	8	461.0	371	1,112	3.00	211	22.4	385.5	2	0.1493	2.18	4.82	2.28	5.32	7.60
69	Roasters	672	2,588	3.85	Nov. 9 to 17	9	934.0	615	2,735	4.45	371	14.3	643.5	2	.1613	2.52	9.05	2.44	6.15	8.59
70	Broilers	800	1,845	2.31	Nov. 10 to 16	7	919.0	761	2,217	2.91	459	24.9	780.5	4	.1682	2.00	8.54	1.86	4.88	6.74
71	Roasters	672	2,741	4.08	Nov. 11 to 18	8	858.0	615	2,660	4.33	143	5.2	643.5	0	.1667	6.00	8.05	5.63	14.64	20.27
72	Springers	400	881	2.20	Nov. 14 to 20	7	462.0	382	1,197	3.13	357	40.5	391.0	2	.1688	1.29	4.28	1.20	3.15	4.35
73	Roasters	672	2,832	4.21	Nov. 13 to 22	10	1,068.0	616	2,862	4.65	217	7.7	644.0	1	.1657	4.92	10.07	4.64	12.00	16.64
74	Springers	400	1,068	2.67	Nov. 16 to 22	7	450.0	389	1,275	3.28	236	22.1	394.5	2	.1630	1.91	4.32	1.83	4.66	6.49
75	Roasters	1,008	4,135	4.10	Nov. 17 to 24	7	1,277.0	954	4,346	4.56	397	9.6	981.0	3	.1627	3.22	12.27	3.09	7.86	10.95
76	Springers	400	1,037	2.59	Nov. 18 to 25	8	490.0	387	1,148	2.97	143	13.8	393.5	1	.1556	3.43	4.92	3.44	8.37	11.81
77	do	400	989	2.47	Nov. 20 to 27	8	486.0	382	1,135	2.97	211	21.3	391.0	2	.1552	2.30	4.89	2.32	5.61	7.93
78	Roasters	672	2,726	4.06	Nov. 20 to 27	8	811.0	635	2,881	4.54	298	10.9	653.5	2	.1552	2.72	8.17	2.74	6.64	9.38
79	Springers	400	1,065	2.66	Nov. 22 to 28	7	422.0	392	1,200	3.06	154	14.5	396.0	0	.1523	2.74	4.33	2.81	6.69	9.50
80	Roasters	672	2,703	4.16	Nov. 23 to 29	7	721.0	647	2,880	4.45	162	5.8	659.5	1	.1560	4.45	7.22	4.46	10.86	15.32
81	do	672	2,698	4.01	Nov. 23 to 29	7	722.0	650	2,755	4.24	143	5.3	661.0	0	.1560	5.05	7.23	5.06	12.32	17.38
82	do	672	2,694	4.01	Nov. 27 to Dec. 2	6	603.0	602	2,705	4.49	182	6.8	637.0	0	.1578	3.31	5.97	3.28	8.08	11.36
83	Springers	672	2,486	3.70	Nov. 29 to Dec. 5	7	737.0	651	2,616	4.02	245	9.9	661.5	2	.1591	3.01	7.24	2.96	7.34	10.30

Fig. 1.—Feeding Station No. 2.

Fig. 2.—Feeding Station No. 3.

Fig. 3.—Feeding Station No. 4.

Fig. 4.—Combination Creamery and Poultry-Feeding Station (Station No. 5).

Fig. 1.—Portable Feeding Battery; Side View.

Fig. 2.—Portable Feeding Battery; End View.

Fig. 3.—Turkey-Feeding Battery.

Fig. 4.—Two Types of Feed Pails.

Fig. 1.—Portable Feeding and Mixing Tank. Note Feeding Pail.

Fig. 2.—Portable Truck for Moving Birds.

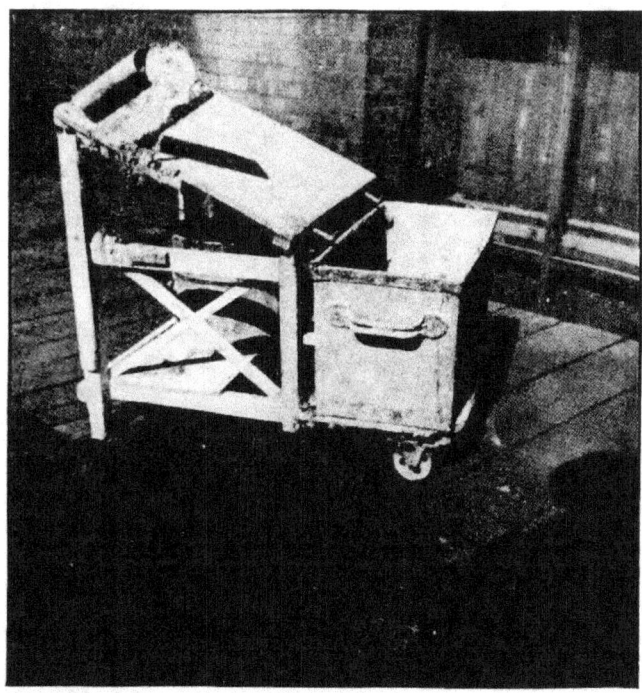

Fig. 3.—Manure Truck.

FIG. 1.—SHIPPING CRATE FOR LIVE POULTRY.

FIG. 2.—ORDINARY LIVE-STOCK CAR, OFTEN USED FOR SHIPPING POULTRY.

FIG. 3.—LIVE-POULTRY CARS, GENERALLY USED FOR LONG-DISTANCE HAULS.

www.ingramcontent.com/pod-product-compliance
Lightning Source LLC
Chambersburg PA
CBHW062335220526
45469CB00008B/2728